PLANT NAMES OF MEDIEVAL ENGLAND

PLANT NAMES
OF
MEDIEVAL ENGLAND

Tony Hunt

D.S.BREWER

First published 1989
D. S. Brewer, Cambridge

Transferred to digital printing

ISBN 978-0-85991-273-0

D. S. Brewer is an imprint of Boydell & Brewer Ltd
PO Box 9, Woodbridge, Suffolk IP12 3DF, UK
and of Boydell & Brewer Inc.
668 Mt Hope Avenue, Rochester, NY 14620, USA
website: www.boydellandbrewer.com

A CiP catalogue record for this book is available
from the British Library

This publication is printed on acid-free paper

To the Memory of my Father

CONTENTS

PREFACE

> A complete language of meaningless nonsense, almost impossible to retain, and certainly worse than useless when remembered – a vast vocabulary of names many of which signify that which is false, and most of which mean nothing at all.[1]

Here is a challenge to the linguist! And it is as a linguist, not a botanist, that I have invaded this waste land where 'full many a flower is born to blush unseen / and waste its sweetness on the desert air'. In the absence of the plants themselves or of their illustrations almost all our deductions about medieval botany are necessarily derived from linguistic evidence and it is the techniques of philology which alone are capable of bringing some sense and shape to the Protean mass of medieval plant names. The apparent anarchy of pre-Linnaean nomenclature does not mean that it had no sense. Yet the elucidation of its sense is a task which has been almost entirely shunned by philologists. Between Bierbaumer's admirable and comprehensive survey of the Old English botanical lexis[2] and various dictionaries of plant names of the modern period[3] there is a conspicuous and deleterious gap embracing the whole of the Middle English period 1100–1500, an age which laid the foundations of vernacular botany and prepared for the work of the great herbalists of the sixteenth century, Turner (c.1508–68), Lyte (c.1529–1607) and Gerarde (1545–1612). Until their debt to medieval botanical sources has been adequately investigated any project for English plant name research which takes the sixteenth century as its starting point is likely to be unreliable.[4] Yet an obvious defence is that the OED is so deficient in botanical materials from the earlier period. Whilst the *Middle English Dictionary* marks a great improvement in this regard, it is based largely on literary evidence and those few English herbals which have secured the attention of scholars, notably the Middle English Macer and the Agnus Castus.[5] These texts are essentially monolingual and contribute less than might be

[1] H. C. Richter in *Hardwicke's Science Gossip* of Nov.1. 1868, p. 251, also quoted in Britten & Holland, *A Dictionary of English Plant-Names* (London, 1886), p. v.

[2] P. Bierbaumer, *Der botanische Wortschatz des Altenglischen* 1–3, Grazer Beiträge zur englischen Philologie 1–3 (Bern / Frankf. am M., 1975–79)

[3] See, for example, R. C. A. Prior, *On the Popular Names of British Plants* (London, 1863; 1870²; 1877³) and Britten & Holland, *op. cit.*

[4] See M. Rydén, 'English Plant Name Research' in S. Johansson & B. Tysdahl (eds.), *Papers from the First Nordic Conference for English Studies, Oslo, 17–19 September 1980* (Oslo, 1981), pp. 374–84.

[5] See G. Brodin (ed.), *Agnus Castus: A Middle English Herbal* (Upsala etc., 1950); G. Frisk (ed.), *A Middle English Translation of Macer Floridus De viribus herbarum* (Upsala etc., 1949). A singularly uninteresting text has been published by P. Grymonprez, *'Here men may se the vertues off herbes' : A Middle English Herbal* (MS Bodley 483, ff. 57r–67v), Scripta 3 (Brussel, 1981). There are many important herbals which await editions, notably those in MSS London, B.L. Arundel 42 ff. 3r–92r; Sloane 5 ff.13ra–57ra; Cambridge,

hoped to our knowledge of the trilingual vocabulary which represents the true reservoir of the plant names of medieval England. This great lacuna must be filled, but with what?

Until the present, the unique remedy lay in the medico-botanical glossary known as the *Alphita*, first published by Salvatore de Renzi from MS Paris, Bibl. Maz. 6954 (with variants from 6957) and later by J. L. G. Mowat, in 1887.[6] But these editions do nothing more than reproduce the texts of arbitrarily chosen MSS (of the fifteenth century!), without the slightest attempt to provide a critical foundation. Whether the fixed identity suggested by a title ('Alphita' is simply the first word of the glossary in many MSS) is real or not can only be known when a thorough study has been made of the work's transmission in all the known MSS. It was while pondering, and recoiling from, the magnitude of this task that I became aware that many of the entries in Mowat's *Alphita* are already to be found in earlier lists of plant names – *synonyma herbarum* – which are so frequently inserted in medical MSS (I have inspected over sixty) to facilitate identification of the ingredients in the receipts which such MSS contain. Here, then, was a new source of materials with which to supplement our meagre knowledge of Middle English plant names (and to add some 500 names to those recorded in the OED, see below pp. xlvi ff). Since I had already embarked on an edition of Anglo-Norman medical receipts,[7] the lists of *synonyma* became of no less practical importance to me than to the medieval *medicus*.

I have concentrated in this study on medieval England, but, of course, Ireland and Wales should not be forgotten. MS London, B.L., Add. 15236 includes a number of Irish plant names which I have published elsewhere[8] and in the same collection, MS 14912 includes a Latin-Welsh vocabulary of plant names which was published by Whitley Stokes.[9] Indeed, for Wales we have medical MSS in the vernacular as early as the fourteenth century and some of them contain receipts which

Trinity College R. 14. 32 ff.1v-65r. All three were used very selectively and idiosyncratically by the compilers of the OED.

6 S. de Renzi, *Collectio Salernitana* 3 (Napoli, 1854), pp. 271–322 and J. L. G. Mowat (ed.), *Alphita: A Medico-Botanical Glossary from the Bodleian Manuscript Selden B.35*, Anecdota Oxoniensia, Med. & Mod. Ser. I, 2 (Oxford, 1887). For an attempted identification of the plants see R. Creutz, 'Das mittelalterliche medizinisch-botanische Vokabularium "Alphita" ', *Quellen und Studien zur Geschichte der Naturwissenschaft und der Medizin* 7, iv–v (1940), 1–80 (437–516 of the continuous pagination). Creutz lists 1210 plant names.

7 *Popular Medicine in Thirteenth-Century England* (forthcoming, Boydell & Brewer). The importance of medical literature as a source of botanical terminology had already been clearly indicated by W. Rothwell, 'Medical and Botanical Terminology from Anglo-Norman Sources', *Zeitschrift für französische Sprache und Literatur* 86 (1976), 221–60.

8 'The Botanical Glossaries in MS London, B.L. Add. 15236', *Pluteus* 4 (1986). Cf. A. Zettersten, *The Virtues of Herbs in the Loscombe Manuscript: A Contribution to Anglo-Irish Language and Literature* (Lund, 1967). See also the very interesting list of herbs in MS London, B.L. Add. 15403 (s.xv) ff.1r–72v.

9 'A List of Welsh Plant Names', *Archiv für celtische Lexikographie* 1 (1900), 37–49. For a Gaelic *synonyma* list ('Aron barba . . .') see J. Mackechnie, *Catalogue of Gaelic Manuscripts in Selected Libraries in Great Britain and Ireland* 1 (Cambr. Mass., 1973), pp.129–34. See also N. Ker, *Medieval Manuscripts in British Libraries* 3 (1983), pp. 456f.

correspond to those in Middle English collections and which naturally purvey a variety of botanical terms.[10]

The present study, whilst being the first of its kind, is essentially a collection of unpublished materials which has been accorded such critical treatment as the present state of our ignorance permits. In presenting over 1800 vernacular names, covering over 600 plant species,[11] I rapidly decided that the most useful way of making the material available was to adopt the form of a dictionary in which the vernacular terms would be listed – as they are in the MSS – under alphabetically arranged Latin headwords. The dictionary form, however, must not deceive. To give the impression that we are dealing with hard facts and certain knowledge would be the opposite of my intention. In the pages which follow I have made liberal use of the question mark as a reminder that my interpretation of the materials is necessarily tentative and as a symbolic indication of my belief that true knowledge is the understanding of one's ignorance. Only with the accumulation of more materials for comparative analysis will the question marks gradually cede their places. The principle, meanwhile, to which I have sought to adhere in recording and identifying the plants is that critical treatment should never be allowed to obscure the data provided in the MSS. Where such treatment is applied it is derived from the widest possible consultation of printed material, relevant MSS, dialect studies etc. and is always explicitly indicated. My main concern has been to establish a satisfactory base for further research and to sift the knowable from the unknowable. The results must not be confused with the stipulative descriptions of a dictionary proper.

Botanical imagery lends itself remarkably well to the description of the task I set myself. All too often it seemed that I was wandering, linguistically speaking, in 'an even mead' that 'nothing teems / but hateful docks, rough thistles, kecksies, burs, / losing both beauty and utility'. Yet all lexicologists know the peculiar pleasure to be derived from studying 'words more sweet, and yet more dangerous / than baits to fish or honey-stalks to sheep, / when as the one is wounded with the bait, / the other rotted with delicious feed'. I have doubtless been caught on a bait of my own making from time to time, but for 'delicious feed' I count myself fortunate to have enjoyed the facilities of the Wellcome Historical Medical Library and the Warburg Institute which remain, unlike certain grander institutions, just what a scholarly library should be, with staff as efficient as they are friendly. I am also grateful to the custodians of the MSS I have consulted for the readiness with which they have made these available to me. My greatest debt, however, is to the British Academy, and particularly Sir Randolph Quirk and Professor M. Screech, for the sympathy with which my plea for more lexical research was received and the

10 See Morfydd E. Owen, 'Meddygon Myddfai: A Preliminary Survey of Some Medieval Medical Writing in Welsh', *Studia Celtica* X/XI (1975–76), 210–33.

11 Whilst Turner is credited with recording about 300 British species, Gerarde is said to refer to about 500, see M. Rydén, 'The English Plant Names in Gerard's Herball (1597)' in *id.* & L. A. Björk (eds.), *Studies in English Philology, Linguistics and Literature presented to Alarik Rynell 7 March 1978*, Stockholm Studies in English 46 (Stockholm, 1978), pp.142–50. Bierbaumer's third volume, which presents the material of the Old English glossaries, has 950 vernacular headwords.

much-needed encouragement which their generous award of a Research Readership afforded me. Without the Academy's support the present work, and much else besides, could not have been undertaken. In a climate so hostile to scholarship, my publisher Dr Richard Barber deserves my best thanks for his enthusiastic interest in research and the practical assistance which he affords to its practitioners. To the University of St Andrews I am grateful for its support of my research over many years and in particular, I wish to thank the Committee on Research in the Arts and Divinity for considerately making a grant in aid of publication. Last, but not least, I should like to thank Mrs Joan M. Jordan for her patience and skill in converting a pile of filing-cards into print without making a single error in the plant names.

INTRODUCTION

The Arrangement of the Materials

The problems which arise from any attempt to record and identify medieval plant names are best illustrated by an explanation of how the materials forming the basis of the present study have been set out.

Latin Headwords: In the case of single sources the headword is given in the form in which it appears in the MS. Even if it is thought to be corrupt, this form is retained and the 'correct' (as determined by the authorities) form is printed afterwards in square brackets, thus: *acumusca* [for acus muscata?]; *testiculus auris* [for muris]. In the case of multiple witnesses, the form of the headword chosen is the one closest to the 'standard' (as determined by the authorities), in many cases identical to it of course. There follow within square brackets a selection of variant forms most of which are taken from the entries containing vernacular items, since I did not attempt to record all the monolingual entries in the *synonyma* lists. There are many unfamiliar Latin words standing as headwords and according to my principle of not obscuring the data I have been conservative in my handling of them. In some cases my choice of the form to stand as headword will seem arbitrary, though I have always used such etymological knowledge as was at my disposition. Occasionally I have preferred a less 'correct' form because it was the almost universally transmitted form in the *synonyma* lists and to obscure this for the sake of a dubious consistency seemed unprofitable.

Botanical Identification: This is obviously the most hazardous part of the study. The sources used here pose a particular problem. Whilst they have the advantage of being lists made up for practical purposes rather than slavish copies of alien treatises, they do not as a rule (they offer few plant descriptions) provide any sort of context which would facilitate identification. In other words, they stubbornly remain lists of names rather than of things. A comparative analysis of all the *synonyma* controlled by such botanical data as is available from other sources (medical receipts, herbals, poetry) [1] does, however, make reasonable progress possible. Where there is agreement between the independent identification of both headword (or lemma) and vernacular synonym (or gloss) I have printed the identification with some confidence. When there was no such corroboration, I have proceeded as follows. Where the meaning of the Latin headword seemed uncertain

[1] Cf. J. Stannard, 'Magiferous Plants and Magic in Medieval Medical Botany', *Maryland Historian* 8 (1977), 33–46 and *id.*, 'Botanical Data and Late Mediaeval Rezeptliteratur' in G. Keil (ed.), *Fachprosastudien: Beiträge zur mittelalterlichen Wissenschafts-und Geistesgeschichte* (Berlin, 1982), pp. 371–95. See also above, p. x, n. 7, for the evidence of plant remains from archaeological sites.

and that of the vernacular synonym clear enough, I have printed the indentification(s) suggested by the vernacular word(s) whilst preceding it/them by a monitory question mark. Where the meaning of the Latin is well established, but the vernacular uncertain, I have adopted a similar procedure, printing the identification(s) obtained from the Latin, again accompanied by a question mark. Where a divergence between the attested meanings of the Latin and the vernacular suggests that the 'synonym' provided is erroneous, I have marked it with an asterisk – where there is only a single entry the vernacular item is asterisked and the correct identification of the Latin headword printed within square brackets. Where the meaning of the Latin is unknown and the vernacular entry susceptible of multiple identifications, I have preferred to print a question mark in place of an identification. Of course, most of the entries include multiple identifications according to the attested meanings of the headword and the variety of synonyms provided. Considerations of space have made it quite impossible to explain and justify the identifications proposed, but I should emphasize that I have based my decisions on as large a corpus of materials as I could assemble. In particular, I have been guided by the accompanying Latin synonyms given in many entries in the MSS but not printed by me (except in rare instances where I felt that they should be recorded) for reasons of space and of purpose – my primary aim being to furnish vernacular plant names. Many of the Latin synonyms are found in the printed text of the *Alphita* and a considerable number recur as headwords. The indices which I have compiled provide a simple cross-reference system. The botanical identifications are made in accordance with the nomenclature adopted in A. R. Clapham, T. G. Tutin and E. F. Warburg, *Flora of the British Isles* 2nd ed. (Cambridge, 1962) – the 3rd edition, 1987, with D. M. Moore replacing E. F. Warburg, appeared too late to be utilized here – and the English names accompanying the botanical identifications are drawn mainly from the same source. It is impossible, of course, to adopt an absolutely consistent principle in the selection of the variously used English names, but I have retained the distinction between genuine English names and invented ones by printing the latter between single quotation marks. I have naturally consulted J. G. Dony *et al.*, *English Names of Wild Flowers: A List recommended by the Botanical Society of the British Isles* (London, 1974). The number of identifications proposed reflects both the instability of pre-Linnaean nomenclature and also the tenuousness of our links with medieval botany. A single plant might be known by a considerable variety of names, often depending on region and locality, and these names in turn used to indicate different plants according to both changes of place and perceived similarities in the properties of the plants. The cases of universal, one-for-one equivalence are rare.[2] Often I attribute no more than the genus (followed by ssp. = (sub)species), since greater precision would be completely arbi-

2 Note the caveats of Clapham *et al.*, *Flora* . . . pp. xiii–xiv: 'English names are frequently only of local use, and they give no reliable information of the relationship or otherwise of the plants, while frequently one name includes a number of distinct entities or is applied to different plants and in various parts of the country. In addition only a small number of plants have English names which are in common use, though many others have names, often translations of the scientific names, which have been given to them mainly by the writers of nineteenth-century floras'.

trary. Despite these restrictions I do not share the radical pessimism of Charles Singer who repeatedly described the attempt to identify Old English plants from their names as 'a fruitless task'.[3] Dialects, the great natural reservoir of archaism, still provide useful evidence and there is enough printed material for comparative purposes to justify, I believe, the present undertaking. However that may be, it remains essential to put into print what materials we have so that the problems may be accurately assessed.

The botanical identifications follow, of course, the now universally accepted system of the Swedish naturalist Carl Linnaeus (1707–78).[4] The application of binomial nomenclature in which taxa are named merely for purposes of reference starts with his *Species plantarum* which appeared in 1753 and covered some 7,700 plants. There are obvious traces of a binary system in pre-Linnaean plant synonymy, a system which goes back to the ancients and Aristotle's method of definition *per genus et differentiam.*[5] Linnaeus's system, which seems essentially to have been the result of his efforts to reform the method of citing books, has one great advantage. With the adoption of a generic name followed by a specific epithet or trivial name, the system achieved a completely new flexibility on account of the fact that the trivial name could be derived from any source, in any mode, and was completely stable and durable – the discovery of new species could be assimilated without any disruption to the existing nomenclature. The genus was based on the form of the flower and the fruit, and generic names, hitherto much more unwieldy, were chosen by Linnaeus for their memorability, being derived from various sources such as classical mythology and even Linnaeus's circle of friends! At first the specific epithet might comprise several words (some even up to twelve!), but it was soon simplified. Most useful of all, it might be drawn from almost any source.

Synonyms: The vernacular names are accompanied by sigla denoting the MSS from which they are derived (for descriptions of the MSS see pp. xix ff below). In the case of pre-1400 MSS (groups A & B) I print all the vernacular entries. In the case of MSS from the fifteenth century (group C) I have been selective, printing all forms that had intrinsic interest, but omitting entries which simply duplicated others. I have admitted a number of Latin phrases or descriptions where they seemed crucially important to the identification of an otherwise obscure name. In the case of MSS which are exclusively Latin-English I have usually omitted the

[3] In J. H. Grattan & Ch. Singer, *Anglo-Saxon Magic and Medicine* (London, 1952), pp. 80–91 and repeated in his introduction to the 1961 reprint of O. Cockayne, *Leechdoms, Wortcunning and Starcraft of Early England*... pp. xl–xlv.

[4] For full accounts see W. T. Stearn, 'Linnaean Classification, Nomenclature and Method' in W. Blunt, *The Compleat Naturalist: A Life of Linnaeus* (London, 1971), pp. 242–49 and J. L. Heller, 'The Early History of Binomial Nomenclature', *Huntia* 1 (1964), 33–70. See also, more generally, W. T. Stearn, 'The Principles of Botanical Nomenclature, their Basis and History', in P. S. Green (ed.), *Plants: Wild and Cultivated* (Hampton, 1973), pp. 86–101.

[5] The most important immediate antecedent of Linnaeus's system is the *Pinax Theatri Botanici* (1623) of Caspar Bauhin. See Stearn, 'The Principles ...'. On changes of nomenclature see A. B. Ashwin, 'Understanding Plant Names and their Changes', *Tuatara* 7 (1958), 84–96.

word *anglice*, but wherever it appeared a useful indication I have retained it. As already explained above, synonyms marked with an asterisk seemed to me *prima facie* to be erroneous (on some of these I shall no doubt be proved wrong). Alterations to the MS readings follow the usual editorial conventions: round brackets for recommended omissions, square brackets for suggested additions. In a very small number of carelessly written words I have silently emended e.g. *c* to *t*. Like Bierbaumer, I have not attempted to arrange the vernacular entries according to the various identifications proposed. My reasons are twofold: first, in many cases this could not be done with any degree of certainty and without a great deal of circular reasoning from existing dictionary entries, and where it could be done with confidence it was largely unnecessary owing to the recognizability of the forms; second, such a procedure would have entirely obscured the chronological sequence of the entries (if only by century) which is itself instructive.

Literature: At the end of many entries I have supplied abbreviated references to a number of works which include relevant discussions or indicate further material. These works are the following:

Agnus Castus: G. Brodin (ed.), *Agnus Castus: A Middle English Herbal*, Essays and Studies on Engish Language and Literature VI (Upsala etc., 1950)

Alph.: J. L. G. Mowat (ed.), *Alphita: A Medico-Botanical Glossary from the Bodleian Manuscript Selden B.35*, Anecdota Oxoniensia, Med. & Mod. Ser. I, 2 (Oxford, 1887) (page and line numbers)

App.: *ibid.*, pp. 201–237 ('Appendix' = glosses of Petrus de Abano)

André I: J. André, *Lexique des termes de botanique en latin* (Paris, 1956)

André II: *id.*, *Les Noms de plantes dans la Rome antique* (Paris, 1985)

B: P. Bierbaumer, *Der botanische Wortschatz des Altenglischen*, 3 vols, Grazer Beiträge zur englischen Philologie 1–3 (Bern / Frankf. am M., 1975–79) (vol. and page numbers)

L: R. J. Stracke, *The Laud Herbal Glossary* (Amsterdam, 1974)

Macer: G. Frisk (ed.), *A Middle English Translation of Macer Floridus De viribus herbarum*, Essays and Studies on English Language and Literature III (Upsala, 1949)

MS: Matthaeus Silvaticus, *Opus pandectarum* (Venice edition of 1498)

N: F. E. S. Norwood, 'The Fourteenth-Century Medico-Botanical Glossary of Raimon de Castelnou', *Proceedings of the Leeds Philosophical and Literary Society* 15 (1972/73), 135–223.

SB: J. L. G. Mowat (ed.), *Sinonoma Bartholomei*, Anecdota Oxoniensia, Med. & Mod. Ser. I, 1 (Oxford, 1882)

The references to the works of Jacques André help to illustrate a particularly important problem, namely the adaptation of classical names to a new flora.[6] The flora described by the Greeks and Romans was southern, representing the plant life

[6] The standard authority on Greek plant names is A. Carnoy, *Dictionnaire étymologique des noms grecs de plantes* (Louvain, 1959). As the latinized Greek names encountered in the present study will be found in the works of J. André, no separate reference to Carnoy seemed necessary.

of the Mediterranean basin and the Near East. The nomenclature inherited by medieval botanists, perforce dependent on the ancient sources, notably Dioscorides, was then applied to the flora of Northwestern Europe.[7] Some of these new applications appear surprising, but it must be remembered that medieval writers had very little first-hand knowledge of the plants which they were reading and writing about. Plants possessing similar properties – leaf shape, colour of flowers, scent etc. – were endlessly confused. Thus the references to André will often reveal considerable divergences between the botanical identity of the name in classical and in medieval usage. Ironically, further confusion was introduced by the glossaries themselves where errors of copying and frequently misunderstood and garbled Greek names led to what might be called 'glossary botany'. Many illustrations will be found in Bierbaumer's discussion of the lemmata of Old English botanical glossaries. Further, I have had the task of accommodating identification to the flora most familiar to insular observers and practitioners. Some of the identifications proposed for Dioscorides or the *Circa instans*, for example, may stand if we assume that the plants were simply imported and that a considerable amount of cultivation took place.[8] However, it seems to me much more likely that those who read and used the somewhat disordered lists of *synonyma herbarum* in insular medical manuscripts identified the plants with the closest *indigenes* or native species and it is among these that I have first sought to identify the plants indicated. All those plants outside of the *flora classica* posed problems to writers on botany, for they were inevitably known by a considerable variety of regional, vernacular names. These names were often subsequently latinized, but sometimes post-classical names (especially Christianizing ones e.g. *herba* + saint's name) were attributed, and at other times existing classical names redistributed. On the one hand, therefore, there was an increasing *décalage* between classical and post-classical nomenclature and this led precisely to the need for synonymies and glossaries. On the other hand, there was always an uncomfortable gap between name and plant, between words and things, and taxonomically there was endless confusion between plants possessing some property in common. Yet, in the case of the *synonyma* –

7 An interesting transitional figure is Marcellus of Bordeaux who in his *Liber de medicamentis* (395–410 AD) refers to approximately 350 plant names and shows considerable, if rather idiosyncratic, interest in *synonyma* and etymology as the basis of plant identification, see J. Stannard, 'Marcellus of Bordeaux and the Beginnings of Medieval Materia Medica', *Pharmacy in History* 15, ii (1973), 47–53. See also *id.*, 'Greco-Roman materia medica in medieval Germany', *Bulletin of the History of Medicine* 46 (1972), 455–68

8 Cf. W. T. Stearn, 'The Origin and Later Development of Cultivated Plants', *Journal of the Royal Horticultural Society* 90 (1965), 279–91, 322–40 & G. Kraus, *Geschichte der Pflanzeinführungen in die europäischen botanischen Gärten* (Leipzig, 1894). L. Voigts, 'Anglo-Saxon Plant Remedies and the Anglo-Saxons', *Isis* 70 (1979), 250–68 argues that the cultivation of southern plants was much easier at the peak of the xerothermic period (c. AD 1000–1200): 'While I do not suggest that the mediterranean plants represented in the herbals grew wild in Anglo-Saxon England, I would argue that it would not have been difficult to cultivate them during the minimal climactic optimum, and there is some scientific evidence of the introduction of plants during this era' (p. 263). See, more generally, H. H. Lamb, 'Our Changing Climate' in F. Perring (ed.) *The Flora of a Changing Britain* (Faringdon, 1974), pp.11–24

which are not simply copies of a single text of remote origins – the situation is perhaps not so desperate. These lists were compiled as practical aids to the understanding and making up of medical prescriptions.[9] Much traditional material, admittedly, is being drawn on and recycled, but a conscious process of selection is always taking place and the medicinal properties of the plants constitute an empirical element. The regularity with which varied lists were compiled and copied suggests that they were found useful. Whatever one may incline to believe about the status of the Old English herbals, these *synonyma* lists were certainly not the product of antiquarianism.[10]

References:

J. Stannard, 'Medieval Reception of Classical Plant Names', *Revue de Synthèse*, sér. gén. 89, 3ième sér., no. 49–52 (1968), 153–62

id., 'Byzantine Botanical Lexicography', *Episteme* 5 (1971), 168–87

C. Vaćzy, 'Nomenclatura Dacică a plantelor la Dioscorides şi Pseudo-Apuleius', *Acta musei Napocensis (Cluj)* 5 (1968), 59–74; 6 (1969), 15–29; 8 (1971), 109–33; 9 (1972), 7–17

R. C. Thompson, 'The Migration of Assyrian Plant-Names into the West', *Classical Review* 38 (1924), 148–9

Indexes: The Index of Vernacular Names is an index of names, not of forms (often of great interest, these must be studied in the relevant entry). It is designed to make the vernacular items more accessible to readers who are not specialists in medieval languages. Plants are therefore indexed under their Modern French (marked by a dagger) and English names where these exist[11] (and, in the case of some obvious compounds, even where they do not) and where the identification is secure. Where there is doubt the Index simply records the name in the form in which it appears in the manuscripts. With the exception of *wild-*, descriptive epithets are usually placed after the noun, which is taken as the headword. The Index of Botanical Names simply records, but does not of course secure, the identification(s) proposed under the Latin headwords. The two indexes are simply a facility which provides some guide to the contents of the present study and must not be treated as an authoritative listing of *synonyma*.

9 See above, p. x, n. 7.

10 See Voigts, *art, cit.*, for a spirited defence of the practicality of the Old English medical texts.

11 The normalized forms of the English items are those which appear as headwords in the OED.

The Manuscripts

A
(1280–1310)

1 London, B.L. Add. 15236 ff. 2r–9r

A collection of botanical, medical and prognostic texts, the first four quires of which (ff.2–52) are probably slightly earlier than the rest of the MS and may be placed c.1300. The list of *synonyma* inc. 'Absinthium galice aloyne, anglice wermot' mixes four languages, denoting the vernacular entries by the terms (usually abbreviated) *anglice*, *gallice* and *hibernice*. The presence of Irish glosses is striking. The list provides little more than synonyms, though there are a few brief descriptions, but these never include degrees and qualities. The entries are economical, often providing no more than one Latin synonym and the vernacular items are evenly distributed between English and French. The small number of Irish glosses I have printed elsewhere.[12] The text is written in long lines, each entry being separated by alternating red and green paragraph marks. It is neatly written, with occasional superscript insertions and a number of blank spaces where the vernacular synonyms have not been inserted. There is no underlining and no annotation. Alphabetical sections are indicated by letters at the top of the page, and the Latin synonyms in each entry are indicated by their initial letter(s) in the outer margin of the page. The section of names beginning with 'v' is preceded by a green capital .V. in the centre of the page (f. 8v). The list of synonyms is an accurate copy of a good model and is closely related to A3 below.

Lit. T. Hunt, 'The Botanical Glossaries in MS London B.L. Add. 15236', *Pluteus* 4 (1986)

2 *Ibid.* ff. 14r–22r

Another list of *synonyma* inc. 'Acacia .i. succus prunellarum agrestium inmaturatarum' which in lay-out and arrangement follows the preceding list, though written in a different hand, yet of the same date. Here the number of vernacular entries is much smaller, whilst the mixture of French, English and Irish is continued. The alphabetical sections are separated by blank spaces. The abbreviations *angl.* and *gall.* are employed where the first list favoured *a.* and *g.* The text is generally reliable. A later hand has added a number of medical receipts and charms at the bottom of ff.14r/v, 15r/v and 16r. The hand responsible for the first list has added entries on six occasions.

Lit. T. Hunt, *art. cit.*

3 *Ibid.* ff. 172v–187v

The list begins with the rubric *Hic incipiunt nomina herbarum una cum quali-*

12 See Lit.

tatibus earumdem et quarundam etiam aliarum rerum cum earum interpretationibus in gallico pro parte et in nostro vulgari. The text commences 'Absinthium amarum *calidum primo gradu, siccum in secundo gradu*, aloyne'. The alphabetical list, slightly later than the two preceding, is written out one entry to a line. The initial letter of each headword is splashed with red and the indication of quality and degree, almost always provided, is underlined in red. Alphabetical sections are marked by alternating red and blue capitals. The Latin synonyms are often indicated in the outer margins by their initial letter(s). The indications *gallice, anglice* are omitted, the French synonyms being given first. There are seven Irish glosses. The text is generally reliable. Folios 182v and 183r are so badly rubbed as to be illegible.

4 *Ibid*. ff. 9r–11v
Two lists of herbs, in the hand and arrangement of A1 above, with rubrics *Ista sunt calida et sicca* inc. 'Asa fetida . . .' and *[I]sta sunt frigida et sicca* inc. 'Arnoglossa . . .', both arranged alphabetically and without descriptions.
Lit. T. Hunt, *art. cit.*

5 *Ibid*. ff. 11v–13v
A miscellaneous list of *synonyma*, without alphabetical order, inc. 'Suffonia a. wodyng [corr. wodeþung], in the hand and arrangement of A1 above. The indications *gallice* and *anglice* are frequently omitted. There are no descriptions. The entries are separated by vertical lines with a red splash, without the paragraph marks employed in A1 above. The text appears to end incomplete at the end of a quire.
Lit. T. Hunt, *art. cit.*

6 London, Lambeth Palace Library 342 ff. 231va–233va
The MS consists of a number of volumes. Following medical receipts, there is a short list of *synonyma*, somewhat carelessly copied, beginning 'Amarisca anglice mawere, unde versus "Fetet amarisca redolet similis camamilla" '. The entries are listed separately without colouring, but with paragraph marks in the ink of the text, and were evidently written at the end of the thirteenth century. The pages are unruled. The list is basically trilingual, though *coluaranis* for *ciclamen* on f.232ra looks Irish. There are few descriptions and no indications of qualities and degrees. The text ends *Expliciunt nomina herbarum* and is followed (f. 233va/vb) by Latin medical receipts.
Lit. M. R. James & C. Jenkins, *A Descriptive Catalogue of the Manuscripts in the Library of Lambeth Palace* (Cambridge, 1930–32), pp. 449–53.

7 Cambridge, Pembroke College 169
There is no modern foliation towards the end of the MS, which is a copy of the *Compendium medicinae* of Gilbertus Anglicus. The recto of the penultimate folio is blank except for the right-hand edge where a late-thirteenth-century hand has added notes on weights and then a single-column miscellany of forty-five plant names accompanied simply by vernacular equivalents, either French or English (rarely both), without the indications *gallice* and *anglice* and with no

descriptions of any sort.
Lit. M. R. James, *A Descriptive Catalogue of the Manuscripts in the Library of Pembroke College, Cambridge* (Cambridge, 1905), pp.163f.

8 London B.L. Sloane 3550 ff. 33ra–36rb
An important medical MS written in a single hand of c.1300. There is a red rubric *De nominibus herbarum* and an explicit *Expliciunt nomina herbarum.* The text, beginning 'Amarusca anglice maythe, "Fetet amarusca redolet similis camomilla" ', comprises a list in two columns in which the separate entries are marked by red splashes on the initial letter of each headword. The alphabetical sections were to have been distinguished by coloured initials, but these were never executed (a few were made good by a later hand). There are some additions in red ink by the hand of the text. The plant descriptions, where they occur, are brief and there is no indication of qualities and degrees. French and English synonyms are prefaced with *gallice/anglice.* An interesting feature of the entries is that for herbs named after saints the designation *sanctus* is consistently omitted, leaving only the first name e.g. *herba Johannis.* The text, which is reliable, is followed by the list of herb substitutes known as *Quid pro quo* (ff. 36va–37va).[13]

9 London, B.L. Sloane 420 ff. 107ra–109va
An important collection of medical texts, including Macer's *De virtutibus herbarum* (ff.145r–179v). The list of *synonyma* is essentially that of A8 above, beginning 'Amarusca anglice mayte, "Fetet amarusca redolet similis camomilla" '. The initial letter of each headword is splashed with red and the alphabetical sections are distinguished by a red capital. The vernacular entries, always introduced by *gallice* or *anglice* (in full), are underlined in the ink of the text. Descriptions are brief and without indications of qualities and degrees. Written in a large, clumsy hand of c.1300, the text is sometimes smudged where the parchment was inadequately prepared and too much ink was absorbed. The same scribe has appended a second glossary (ff.109va–113va) beginning 'Agrestis dicitur aliqua herba quandoque campana quandoque salvatica' which has only two vernacular entries (f.110ra) 'Bedaga vulgari nostro eglenter', (f.110va) 'Corulus arbor gallice coudre'.

10 London, B.L. Sloane 3550 ff. 41r–76r
An extensive medical glossary inc. 'Aer idem quod divus', ending 'Nunc requiem repetam capiunt sinonima metam', but containing only some two dozen vernacular entries (French and English). Frequent use is made of red and blue paragraph marks and of wavy red lines to effect subdivisions of the text. Although the entries are quite short, the contents may range from description, account of origins ('Agrippa est unguentum a rege Agrippa in antidotario'), indications of qualities and degrees, to references to sources ('Sic Girardus', 'ut in libro urinarum Ysaac', 'ut in Alexandro de oculis', 'ut in Theophilo' etc.). The text is very neatly written, c.1300. The alphabetical sections are distin-

[13] See P. Berges, *Quid pro quo: Geschichte der Arzneimittelsubstitution* (Marburg, 1975).

guished by decorated red and blue initials. The scribe has added a number of entries in red. Another copy of this glossary, covering only the letters A–C, is found in MS B.L. Sloane 348 (s.xiii²), ff.1ra–4vb.

11 Oxford, Bodleian Library, Digby 69 ff. 66r–84vb
A text of Macer's *De virtutibus herbarum* with a small number of thirteenth-century vernacular glosses beside the rubrics introducing each plant. The MS also contains a purely Latin set of *synonyma* inc. 'Alphita farina ordei idem' on ff. 92r–94v.

12 Cambridge, Gonville and Caius College 95 (47) pp.1a–3a
A medical MS with a short list of *synonyma*, without colour, plant descriptions, or any strict alphabetical order, beginning 'Amarusca anglice mayþe, versus "Fetet amarusca redolet similis camomilla" ', written at the end of the thirteenth century (not the fifteenth as James inexplicably says in his catalogue). Vernacular entries are marked by *gallice/anglice*. As in A9 above, this list is followed by another (pp. 3a–6a), beginning 'Agrestis dicitur quandoque aliqua herba . . .', with the same two vernacular glosses 'bedagar vulgari nostro eglentere' and 'cosilus [corr. corilus] gallice couder'.
Lit. M. R. James, *A Descriptive Catalogue of the Manuscripts in the Library of Gonville and Caius College* 1 (Cambridge, 1907), pp. 94–96

B
(1310–1400)

1 London, B.L. Sloane 420 ff. 115r–121r
A heavily annotated glossary with the red rubric *Hic incipiunt nomina cinonoma herbarum*, beginning 'Hic incipiunt nomina herbarum primo de arthemesia. Arthemesie tres species sunt . . .' The initial letter of each headword is splashed with yellow. Each entry is rarely anything more than a number of synonyms, the vernacular items being introduced by *g'* or *ā*. Sometimes a blank space has been left where a French synonym was to be provided. A later hand has added grey paragraph marks to distinguish the alphabetical sections. There are almost no descriptions, but reference to qualities is quite frequent. The hand is of the first quarter of the fourteenth century. The volume as a whole is a compendium of various medical MSS put together by Roger Marchall, who died in 1477 (the list of contents on f.1v is in his hand, as is that in MS B.L. Harley 531 f.1*). The list of synonyms has no obvious source. It has received additions in various fourteenth-century hands, and also a considerable number in a fifteenth-century hand (see C20 below).

2 London, B.L. Sloane 5 ff. 4ra–12va
A medical MS which includes an important Middle English herbal (ff.13ra–57ra), apparently translated by John Lelamour of Hereford. The list of *synonyma* is from an earlier MS and has the red rubric *Hic incipiunt cinonima de herbis latine, gallice et anglice*, beginning 'Absintheum amarum deu maners

gallice aloyne anglice weremod'. The headword of each entry is accompanied by a red paragraph mark, and each alphabetical section is marked by a red capital. The entries consist exclusively of synonyms (vernacular items introduced by *g.* and *a.*) and indications of qualities and degrees (heavily abbreviated). There are scarcely any descriptions. The whole glossary is written in grey ink and in a large hand of c.1340.
Lit. G. Frisk (ed.), *A Middle English Translation of Macer Floridus De viribus herbarum* (Upsala, 1949), p.16.

3 London,B.L. Sloane 2479 ff. 90r–107r
To the list of *synonyma* beginning 'Artemesia mogwort' a later hand has added a black rubric *Herbarum nomina lat[i]no-anglica*. The entries are written in a single column on unruled pages. The first letter of each headword is splashed with yellow. The entries are seldom anything but synonyms, often single synonyms. There are spaces between each alphabetical section. Since the glossary is bilingual only, the designation *anglice* is very rarely used. The rest of the MS comprises many fragments of different dates including the *Thesaurus pauperum* of Peter of Spain (ff.10ra–41va) and ME medical receipts (ff.81r–89v). The list of *synonyma* dates from the end of the fourteenth century.

4 London, B.L. Add. 17866 ff. 43ra–45rc
An early-fourteenth-century MS largely composed of Middle English medical receipts in both prose and verse and a verse herbal. The list of *synonyma* has a red rubric *Hic incipiunt nomina herbarum*. The arrangement is disordered. In principle there is an English headword (initial letter in red) and listed below it are synonyms in Latin and English, with a few in French. A peculiarity is that, although the list is divided into alphabetical sections (each preceded by a large red capital), the alphabetical principle really applies not to the English headwords, but to the first synonym given for each entry, which is usually Latin. This method has led to various errors and displacements, and since no other information is provided by the list, its evidence has to be used with caution. It contains a number of unique entries.

5 London, B.L. Harley 2558 ff. 1ra–1vb & 2ra–5vb
A make-up volume which is of particular interest for the history of English medicine. The first glossary is incomplete owing to the loss of leaves from the MS and ends with the entries 'lingua cervi scolopendria idem; lingua avis .i. pigle'. It begins 'Allelluia anglice wdesure'. Each headword is preceded by a red paragraph mark and alphabetical sections are distinguished by red capitals. The entries contain synonyms introduced by *gallice/anglice* and a few short descriptions. The second glossary, in the same hand (s.xiv[1]), begins 'Artemisia herba diane a gentilibus consecrata est . . .' and ends 'Vervena: verveyne'. Here headwords receive a red initial letter and the alphabetical sections are distinguished by blank spaces. Descriptions, where they are provided, are longer than in the first glossary. Throughout the second list there is a number of blank spaces where vernacular items have not been provided, and the designations *gallice* and *anglice* are often dispensed with.

Lit. T. Hunt, 'The "Novele Cirurgerie" in MS London, British Library, Harley 2558", *Z. f. rom. Phil.* 103 (1987) [271–99], 275–79

6 London, B.L. Royal 17 A VIII ff. 75ra–77rb
A collection of Middle English medical receipts copied at the end of the fourteenth century. The list of *synonyma* (red rubric *Nomina arbarum*) is short and incomplete (ends with *mercurialis*, the rest of f. 77rb being left blank). Written in two columns, the list consists of a Latin headword with a single English equivalent below, and begins 'Arthemafia [sic] mugwort'. It rapidly becomes confused, with errors, displacements etc., so that it is of very limited value. No information other than *synonyma* is furnished.
Lit. G. F. Warner & J. P. Gilson, *Catalogue of Western Manuscripts in the Old Royal and King's Collections* 2 (London, 1921), p. 216.

7 London, B.L. Royal 12 G IV ff. 134ra–136vb
Medical receipts and treatises, including the *Compendium medicinae* of Gilbertus Anglicus and a receipt collection by John of Greenborough, make up the bulk of the MS, which consists of pieces copied in the thirteenth, fourteenth and fifteenth centuries. The list of *synonyma* was copied at the end of the fourteenth century and is written in a large hand. The headwords are written larger than the entries and are underlined in red. The text begins 'Aaron barba Aaron jarus pes vituli gallice jare anglice cokousepintul vel ȝekestres calida et sicca in primo gradu'. After the letter A the entries are reduced to little more than *synonyma*, with sporadic indications of qualities and degrees. The alphabetical sections are distinguished by red paragraph marks and blank spaces. Vernacular entries are introduced by *g*^{e} and *a*^{e}.
Lit. G. F. Warner & J. P. Gilson, *Catalogue* . . . 2, pp. 69–71.

8 Cambridge, Gonville and Caius College 451a ff. 15r–16ra
A collection of medical receipts and a herbal. The list is simply a set of Latin headwords in alphabetical order, accompanied by a single vernacular equivalent without the designation *gallice/anglice*. There is no colour or decoration and the list is simply headed *Hic incipiunt nomina herbarum in lingua latini, italici, anglici*. It was written in the second half of the fourteenth century.
Lit. M. R. James, *A Descriptive Catalogue of the Manuscripts in the Library of Gonville and Caius College* 2 (Cambridge, 1908), p. 525.

9 London, Royal College of Physicians 227 ff. 182r–184r
A collection of medical treatises including the *Cirurgia* attributed to Roger Frugardi (ff. 8r–89r). The list of *synonyma* is written in a large, clear hand of the first half of the fourteenth century. The text begins 'Amarusca gallice amaroc anglice maythes . . .' and is written out in long lines, orange slashes being used to separate the entries, and orange capitals to mark the alphabetical sections. The entries are short, with little description. Vernacular synonyms are not always given.
Lit. N. R. Ker, *Medieval Manuscripts in British Libraries* I (Oxford, 1969), pp. 199–201 (read MS 227 not 227a)

C
(1400–1500)[14]

1 London, B.L. Royal 12 E 1 ff. 69r–107v
A volume made up of various MSS from the late thirteenth century to the late fifteenth century, comprising religious and medical texts. The list of *synonyma* contains an exceptionally large number of entries, though they are confined almost entirely to the simple provision of synonyms, in Latin, French and English. The designations *gallice/anglice* are frequently dispensed with. The vernacular material is richer here than in any other glossary. The list begins, without rubric or heading, 'Absinthium amarum anglice wermod'. There is no colour, decoration or annotation. The list bears no clear relation to any of the other *synonyma* lists used in the present study, and is valuable no less for the Latin material than for the vernacular. It was copied at the beginning of the fifteenth century.
Lit. G. F. Warner & J. P. Gilson, *Catalogue* . . . II, pp. 48–50.

2 London, B.L. Sloane 2527 ff. 219r–223v
A make-up volume which includes a thirteenth-century copy of the *Antidotarium Nicolai* (ff. 2r–21v) and a large number of medical treatises and receipts copied in the fifteenth century in both English and Latin (there is a Middle English text of Macer on ff. 251r–293v). The scrappily written glossary has the heading *Sinonoma herbarum in anglico et primo per A literam.* It is distinctive in being an English-Latin vocabulary, the English headwords simply being followed by Latin synonyms. The copy shows serious signs of disorder. There are blank spaces on ff. 221r and 222r; the whole of f. 223v is blank; there is a leap from *h* on f. 222r to *m* on f. 223r and the whole text ends incomplete with 'mistell of an ok latine viscus quercinus'. The Latin synonyms are introduced by *latine* (*l*^e^).
Lit. Frisk, *A Middle English Translation of Macer Floridus* . . . pp. 24–5

3 London, B.L. Sloane 3149 ff. 2r–13r
A slim volume of medical texts. The list of *synonyma* begins, without heading or rubric, 'Atriplex agrestis criselocanna anglice meldys'. Both headwords and English synonyms (there are no French) are underlined in red. The entries provide little more than synonyms, though occasionally there are references to qualities and degrees. The alphabetical sections are distinguished by red capitals and blank spaces. The copy seems to date from the middle of the fifteenth century.

[14] I have not included the single-leaf fragment of a *synonyma* list in MS Cambridge, University Library Dd. 5. 53 (s.xv) f. 2ra/b (inc. 'Arthemasia . . . anglice mugwort'), though I have recorded the one new gloss which it yields, see *adaserta*.

4 London, B.L. Sloane 3545 ff. 5v–11r

A very slim volume of only eleven folios, beginning with *De virtutibus duodecim herbarum*. The list of *synonyma* begins with the rubric *Incipit liber cinonomorum herbarum* inc. 'Aliapiades .i. laureola herba catholica anglice laureall cuius fructus dicitur coconidium'. Both headwords and English synonyms are underlined in red (there are no French entries), the English synonyms mostly being written on the right-hand side of the text. The information provided rarely exceeds the furnishing of synonyms. The copy dates from the second half of the fifteenth century.

5 London. B.L. Sloane 964 ff. 63r–82r

An interesting collection of Middle English medical receipts. The list of *synonyma* begins, without rubric or heading, '[A]aron barba Aaron jarus pes vituli idem sunt gallice jare anglice coccouspyntel vel ȝekesters, calida et sicca in primo gradu'. There is no colouring or decoration. The initials intended to distinguish the alphabetical sections were never inserted. The entries, though variable in length, regularly include descriptions, vernacular synonyms (French and English) and references to qualities and degrees. C5 and C6 (see below) are based on a common original and contain a rich store of valuable lexical material. This copy dates from the second half of the fifteenth century.

6 London, B.L. Sloane 347 ff. 76r–97v

An important medical MS which contains an acephalous copy of John of Arderne's *Practica cirurgiae* (ff. 8r–75v). The list of *synonyma*, very close to that in C5 above, begins, without rubric or heading, '[A]aron barba Aron jarus pes vituli idem sunt gallice jare anglice cokkyspyntel or rampe vel ȝekesters calida sicca in primo gradu'. There is no colouring or decoration. The alphabetical sections are separated by blank spaces, the coloured initials intended to distinguish them never having been executed. At the top of f. 84r is written 'Richardus Hawkusbury electus est in abbatem Eveshamie sexto die junii anno domini m. cccc. lxvii'. The last few entries are missing, probably as a result of the loss of several folios. A number of additions has been made by near-contemporary hands. The copy dates from the second half of the fifteenth century.

7 London, B.L. Harley 3388 ff. 75r–86v

A collection of medical texts. The list of *synonyma* has the rubric *Incipit cynonoma herbarum* and ends with a scroll design with the words 'medicine contra dolorem capitis sunt que sequntur'. The list begins 'Arthemasia domestica .i. febriifuga nostra .i. vetonica feverfu anglice fetherfoye'. The first letter of each headword is splashed with red and the English (but not the French) synonyms are underlined in red. The entries are confined to synonyms and, more occasionally, brief descriptions. Alphabetical sections are distinguished by red capitals and blank spaces. Between A and B (f. 76v) a number of medical receipts has been written. The whole copy has been very heavily annotated by a variety of later hands and a number of corrections made. The copy dates from the second half of the fifteenth century.

8 London, B.L. Arundel 42 ff. 92ra–98rb
At the end of a striking English herbal (see C9 below) there is a list of *synonyma* headed in red *Alphabetum herbarum* inc. 'Aaron barba Aaron jarus pes vituli idem sunt gallice jare anglice cockouwespenteles vel ȝekesters calida et sicca in primo gradu'. The headwords are accompanied by a red paragraph mark, the alphabetical sections are distinguished by blue capitals. The entries include French and English synonyms, occasionally descriptions, and more rarely indications of qualities and degrees. The copy, made in the first half of the fifteenth century, yields a great deal of useful vernacular material.

9 *Ibid.*, ff. 3r–92r
A striking Middle English herbal (unillustrated) which cites a considerable range of authorities. It is incomplete, stopping at *gazarus*, where it is immediately followed by the list of synonyms (see C8 above). The first entry is *abies*. Each entry consists of a list of synonyms (Latin and vernacular), a detailed description of the plant (apparently both first- and second-hand), and receipts in which the plant is used. The synonyms listed are often surprisingly numerous and many items are not attested elsewhere. Rubrics are in red and each plant entry begins with a blue capital. The scribe is the same as that of C8 above.

10 London, B.L. Sloane 282 ff. 167va–173vc
A vast compendium of medical texts, including Macer's *De virtutibus herbarum*, mostly in the hand of a single scribe of the second half of the fifteenth century. The list of *synonyma* begins with a red rubric *Hic incipiunt cinonoma de nominibus herbarum secundum magistrum Johannem Bray* and ends with a red explicit *Explicit liber cynonomum editus a magistro J.B. de nominibus artem phisicalem et sirurgiter tangentibus in genere.* At the top of each recto page is written in red *Synonoma secundum m. Johannem Bray.* Master John Bray (d.1381) was physician to Edward III. The text of the *synonyma* is written in three columns (two on f.167v) and begins 'Alleluya panis cuculi . . .' Each alphabetical section is marked with a red capital. The synonyms (French, English and Latin) are bracketed together in each entry. Little use is made of the designation *gallice/anglice.* Where there is a short description, this is always in English. There are no references to qualities or degrees. The scribe has made a number of additions at the bottom of the page. The glossary is essentially the same as C11 below.
Lit. H. E. Ussery, *Chaucer's Physician: Medicine and Literature in Fourteenth-Century England* (New Orleans, 1971), p. 42.

11 Cambridge, University Library Dd. 10. 44 ff. 104ra–112rc
A medical MS which includes the *Practica brevis* of Johannes Platearius in an English translation (ff.1–100). The list of synonyms begins 'Alleluya panis cuculi payn de cokyll wodsoure stubwort' and ends *Explicit sinomorum.* The text is written in three columns, each headword being splashed with red and accompanied by a paragraph mark, the synonyms then being simply listed. The alphabetical sections are distinguished by alternating red and blue capitals. As in C10 above, to which it is closely related, the few descriptions are always in English.

There are two sections covering 'B', occurring on either side of the section 'C'. *Lit. A Catalogue of the Manuscripts preserved in the Library of the University of Cambridge* 1 (Cambridge, 1856; repr. 1980), pp. 433–35.

12 London, B.L. Add. 18752 ff. 93r–117r

A miscellaneous volume of MSS and incunabula from the fourteenth to the sixteenth century which includes astrological and medical texts. The list of *synonyma* begins with the red rubric *Hic incipiunt sinonoma et equivoca diversarum specierum herbarum, gummarum et arborum litteratim per litteras alphabeti* (repeated as explicit). At the top of f. 93r is written the name Johannes Frostelle (see also Frostelle at the top of f.123v). The list begins 'Arthemesia mater herbarum matricaria vel marcha anglice mogwort calida et humida in tertio gradu'. On ff. 93r–95r and 109r–112r headwords and vernacular synonyms are underlined in red. There are frequent cross-references in red of the type 'requerre in c et s'. The alphabetical sections are separated by large blank spaces – only rarely have the intended red capitals been executed. There are a few descriptions and sporadic references to qualities and degrees. The synonyms are introduced by *gallice* and *anglice*. The treatise known as *Quid pro quo* is found on ff.119v–122r.
Lit. Catalogue of Additions to the Manuscripts in the British Museum MDCCCXLVIII–MDCCCLIII (London, 1868), pp.145f.

13 London, B.L. Sloane 962 ff. 252r–259rb

A large compendium of medical treatises and receipts in Latin, French and English, including Macer's *De virtutibus herbarum* (ff. 90r–123v). The list of *synonyma* has been added to the compilation from another (paper) MS. It begins, without rubric or heading, 'Aaron barba Aaron jarus pes vituli jare cokkowyspyntil ȝekesters satirion'. The text is written in single columns as far as f. 256v and in double columns from ff. 257ra to 259rb. The last folio is badly mutilated, only a few entries surviving. The loss of folios in the middle of the MS has led to the disappearance of all entries from D to R inclusive. It is clear that the present list represents the conflation of three distinct MS fragments. In the first (ff. 252r–256v) the entries are little more than strings of *synonyma*. Little use is made of the designations *gallice* and *anglice*. There are few descriptions and no references to qualities and degrees. The alphabetical sections are distinguished by red capitals and red letters at the top of each page. The hand is that of the second half of the fifteenth century. On ff. 254r and 255r a slightly later hand has added descriptions of *burneta, bursa pastoris, birula, branca ursina, borago silvestris*. The second fragment is written in a slightly larger hand, in two columns, and begins 'sapus saracenicus fert de forti lexivo et sepo et oleo . . .' and finishes (f. 257vb) 'Explicit sinomia secundum Raysym doctorem phisice'. It is exclusively Latin and deals also with medical terms. The third fragment, in a smaller hand, begins (f. 258ra) 'Salmatina [corr. sal marina?] aqua maris' and continues to f. 259rb which is mutilated. There are English synonyms but the designation *anglice* is dispensed with. Another list of *synonyma* was apparently begun on f. 259v, but only a few entries survive the mutilation of the folio.

14 *Ibid.*, ff. 10v–11v
A small list of *synonyma* written as consecutive text in widely spaced alphabetical sections, the initial letters of which are splashed in red. There is a red rubric *Nomina herbarum in latinum* and the text begins 'Antemus .i. camamille'. The entries are exclusively *synonyma* in Latin or English (no use is made of *anglice*) and their number barely exceeds one hundred. The hand is of the second half of the fifteenth century.

15 *Ibid.*, ff. 87r–89v
The first page of this list is written with double spacing, the vernacular synonyms being inserted as interlinear glosses to the simple succession of headwords, the initial letters of which are splashed with red. Thereafter the text, which begins 'Marubium horhowne', is written as consecutive prose without colour, underlining or paragraph marks. The vernacular *synonyma* are introduced by *gallice/anglice*. There are sometimes brief descriptions. On f. 87r and the upper third of f. 87v there is no alphabetical arrangement. With the entry *anabulla* a third of the way down f. 87v alphabetical order is maintained until 'Cristiana habet folia rotunda' on f. 89v. Then a new sequence *arnoglossa – elatericum* occupies the rest of the last page. The hand is of the second half of the fifteenth century.

16 London, B.L. Sloane 405 ff. 7r–17v
A volume made up of two MSS comprising medical treatises and receipts in Latin and English. The list of *synonyma* begins, without rubric or heading, 'Absinthium amarum idem gallice aloyne anglice wormode'. The entries are composed almost wholly of synonyms (Latin, French and English), but there are some references to qualities and degrees. The alphabetical sections are distinguished by red capitals. The initial letter of each headword is splashed with red. There is a number of additions in the hand of the text, which dates from the second half of the fifteenth century.

17 London, B.L. Sloane 962 f. 155r/v
A list of plants under the heading *Yceux sunt que engendrount let* in which the Latin names are accompanied by their vernacular equivalents introduced by the formula 'c'est . . .'

18 Oxford, Bodleian Library, Digby 29 ff. 38r–44v
A large compendium of medical treatises and receipts in Latin, French and English. The list of *synonyma* begins with a red rubric *Hec sunt nomina synonyma herbarum* inc. 'Absinthium amarum idem gallice aloyne anglice weremode'. The entries are written in long lines, the initial letter of each headword being written in red. There are references to qualities and degrees, short descriptions, and the vernacular entries are introduced by *gallice* (*g.*) and *anglice* (*a.*).
Lit. G. D. Macray, *Catalogi codicum manuscriptorum bibliothecae Bodleianae*, pars IX (Oxonii, 1883), cols. 25–28.

19 London, B.L. Sloane 1067 ff. 280va–288ra
A vast compendium of medical treatises including the *Rosa Medicinae* of John of Gaddesden (ff.1r–280r). The list of *synonyma* begins, without rubric or heading, 'Aaron pes vituli pes iudei idem'. It is written in two columns, the alphabetical sections being distinguished by blue initials decorated with red. The entries are short, normally confined to synonyms (Latin, French and English). The vocabulary is notable for the inclusion of many Arabic words. It is followed by the list of plant substitutes known as *Quid pro quo* (ff. 288ra–289rb) and by a text of the *Alphita* (ff. 289rb–303rb).

20 London, B.L. Sloane 420 ff. 115r–121r
The additions in a fifteenth-century hand to B1 above.

21 Cambridge, University Library Dd. 11. 45 ff. 94r–116va
A compilation of medical treatises including the *Thesaurus pauperum* of Peter of Spain (ff.1r–62v). The list of *synonyma* begins, without rubric or heading, '[A]marusca mayes'. It is written in double columns, many parts of which have been left blank (ff.100r/v and 102v are completely blank). The entries are mostly confined to single English equivalents (*anglice* is scarcely ever used), but often no English synonyms are provided at all. There is no colour or decoration. This lengthy glossary, which is not entirely reliable, appears to be a conflation of various sources, hence the blank spaces that were intended to be filled and the duplication of quite a few entries.
Lit. A Catalogue of the Manuscripts preserved in the Library of the University of Cambridge 1 (Cambridge, 1856; repr. 1980), pp. 445–46.

22 Exeter, Cathedral Library 3519 ff. 52ra–61vb
A compilation of medical works including many receipts and a text of Macer's *De virtutibus herbarum* (ff. 2r–40r). The list of *synonyma* begins without rubric or heading, 'Absinthium amarum calidum primo gradu, siccum in secundo, aloyn, wermot'. There is a coloured initial for each entry. The entries include synonyms (Latin, French and English; little use is made of *anglice*) and references to qualities and degrees. There are some very brief descriptions. The alphabetical sections become shorter as the text proceeds.
Lit. N. R. Ker, *Medieval Manuscripts in British Libraries* 2 (Oxford, 1977), pp. 828–30.

23 Cambridge, Gonville and Caius College 200 (106) pp.196a–214
A miscellany of religious, herbal and medical texts. The list of *synonyma* begins, without rubric or heading, 'Absinthium amarum deu maneris de aloine anglice wermod'. The initial letter of each headword is splashed with red and the alphabetical sections are marked with a red or blue capital. The entries are mainly confined to synonyms (Latin, French, English; *gallice/anglice* not always used). A new hand starts on p. 210b. Page 213 is not divided into two columns. Page 214 is in the hand which is responsible for a number of marginal additions to the text: two thirds of the page are taken up by very short Latin plant descriptions. The copy is careless and inaccurate with frequent errors and

displacements of material.
Lit. M. R. James, *A Descriptive Catalogue of the Manuscripts in the Library of Gonville and Caius College* 1 (Cambridge, 1907), pp. 229–30.

24 Cambridge, St. John's College E 6 (109) ff. 70ra–83va
A selection of religious, prognostic and medical texts. The list of *synonyma* begins, without rubric or heading, 'Alleluya wodesour chemerek stevenwort' and ends 'Explicit herbarium'. The text is written in two columns, the headwords being written larger than the main text and frequently underlined. The Latin and English synonyms are then listed below them (without the designations *anglice/gallice*). There are a few descriptions in English.
Lit. M. R. James, *A Descriptive Catalogue of the Manuscripts in the Library of St. John's College, Cambridge* (Cambridge, 1913), pp.142–3

25 Cambridge, Gonville and Caius College 198 (104) pp. 85a–86a
A copy of the *Antidotarium Nicolai* followed by a short vocabulary written in two columns with headword and vernacular (English) synonym well separated, giving the appearance of five columns. There is no colour, the opening initial never having been inserted. The text begins '[A]cardella sowþystell'.
Lit. James, *Catalogue*... 1, pp. 226–7.

26 London, B.L. Add. 27582 ff. 31r–68v
A large compendium of medical MSS and incunabula in Latin and English. Folios 31r–68v are occupied by a vocabulary in which a well-spaced set of alphabetically arranged Latin synonyms are bracketed to a single English word. The list includes stones, metals etc. as well as plants. It is written in two columns, each group of synonyms being numbered within the appropriate alphabetical section (which is marked with a red capital) and bracketed to the English headword with red lines. Thus 'A' contains 265 entries, 'B' 50, 'C' 79, thereafter diminishing as the work proceeds. On ff. 2ra–30va there is a comprehensive index and an inscription 'Hunc librum do et lego domino Arnulpho ap Rice suppriori sancti Johannis Exonie ad orandum pro anima mea videlicet Ricardi Smyȝth anno domini 1523'.
Lit. Catalogue of Additions to Manuscripts in the British Museum in the years MDCCCLIV–MDCCCLXXV vol. 2 (London, 1877; repr. 1967), pp. 332–3.

27 Cambridge, University Library Dd. 11. 45 ff. 145r–153v
The list of *synonyma* is written in a different hand from the rest of the MS. The text begins 'Alleluya payn de cuculle wodsowr stubwort' and the first six entries are written in red. On f.145r/v the text is written out in long lines with the initial letter of each entry in red, but from f.146r it is written in single columns alternating red and black (ink of the text) initials. The list ends incomplete at *quinquenervia*. The entries consist solely of Latin and English (occasionally French) synonyms without the designation *gallice/anglice*.

28 London, Society of Antiquaries 101 ff. 88ra–90ra
A collection of miscellaneous texts including medical receipts. The list of *sy-*

nonyma, which is disordered, is written in three columns (two on f. 90r) and there is no colour or decoration. Folios 88ra–89rb offer an alphabetical glossary of single-word (Latin-English) *synonyma* from 'Arthemesia mugworte' to *titimallus*. The rest is taken up by descriptive entries which include references to qualities and degrees, without proper alphabetical order. On f. 89va a new set of entries starts with *acetosa*. Only very occasionally is the designation *anglice* used. The first list is a rather careless copy of an interesting model, including some Old English words.
Lit. N. R. Ker, *Medieval Manuscripts in British Libraries* 1 (Oxford, 1969), pp. 304–6.

29 Oxford, Bodleian Library, Ashmole 1389 pp. 268a–279a
The list of *synonyma* is the second MS, on parchment (the first, on paper, is a collection of medical receipts), and begins 'Alfita farina ordii'. There are red initials for each alphabetical section and red splashes on the first letter of each headword. The list consists exclusively of synonyms and in English only. Some parts (e.g. p. 270) are badly rubbed and have been wrinkled by damp.
Lit. W. H. Black, *A Descriptive, Analytical and Critical Catalogue of the Manuscripts bequeathed unto the University of Oxford by Elias Ashmole M.D., F.R.S* . . . (Oxford, 1845), cols. 1076–78.

30 Oxford, Bodleian Library, Ashmole 1397 pt. 2, pp. 30–40
The volume is composed of two MSS, the first dating from the end of the fourteenth century, the second from a slightly later date. Both contain medical texts. The list of *synonyma* is written in black ink and the headwords are underlined in red. The entries include articles in which the descriptions (qualities, degrees etc.) are entirely in English and cover minerals and precious stones as well as plants. Each alphabetical section is separated from the next by a red line. There are a few entries which have been squeezed into the space at the bottom of p. 29 after the explicit of the *Quid pro quo* (pp. 25–29) and include 'eupatorium, wild nepe'. The list, which is not primarily a collection of *synonyma*, begins 'Abrotanum gallice averoigne'. The *Quid pro quo* has the following rubric: *Hic incipit tractatus qui vocatur Quid pro: yt is on englysch because þat all diverse medicynus may not be founde in dyverse countres. Here is set o þing for a oþur.* It includes vernacular entries like 'pro aspalto: tar or skyrewhitsede' and 'pro soþernwod: faitourwort or wormod'.
Lit. Black, *Catalogue* . . . cols.1090–94.

31 Oxford, Bodleian Library, Ashmole 1443 pp. 87–189
An English herbal, rather than a collection of *synonyma*, very neatly written in the centre of the page and beginning 'Absinthium wormewode aloyne symount'. The colophon on pp.189–90 runs: 'Here endiþ tþis herbary þe whiche ys draw out of þe nw Circa Instans, and hyt towcherþ schortlyche þe principal vertuys and þe special effectis of herbis and droggis þat beþ most comyne in use and her dyvers grees of qualites or of her complexions and her propur and most special kynd of worcheyng. Amen'. The herbal is written on paper, alphabetical sections are separated by blank pages, the entries are written consecu-

tively, not in a column, with the headwords distinguished by red paragraph marks and with occasional red rubrics. The entries frequently include not only *synonyma*, but also the medicinal properties and uses of each plant. Pages 169 and 170 are mutilated.
Lit. Black, *Catalogue* . . . cols.1202–5.

32 Oxford, Bodleian Library, Ashmole 1447 pp. 205–16
Part of a volume containing eight MSS of different dates, all purveying medical texts. There is a simple, rather scrappy list of *synonyma*, composed of short entries, beginning 'Arthimesia .i. mater herbarum .i. moderwort, mougwort'. It contains many obvious errors and is heavily annotated, sometimes with English glosses. There is an occasional French word preceded by *gallice*. The first letter of each headword is splashed with red.
Lit. Black, *Catalogue* . . . cols. 1220–26; Brodin (ed.), *Agnus Castus*, pp. 90f.

33 Oxford, Bodleian Library, Ashmole 1477 pp.100–110
One of twelve paper MSS which compose this volume of medical texts, the list consists of single-line entries, with red splashes marking the first letter of each headword, beginning 'Antimomum is hote and drye in the iiiith degre . . .' In the margins there is a mixture of headwords and English equivalents written in red. Occasionally English *synonyma* are found in the body of the text written in red or brown. The descriptions are entirely in English.
Lit. Black, *Catalogue* . . . cols. 1293–99.

34 Oxford, Bodleian Library, Ashmole 1477 pt. 2 ff. 29r–33v
A very scrappy list, written in a large hand, beginning with a red rubric *Hic incipiunt nomina herbarum secundum Galienum, Socratem, Alexandrum et Ypocratem* and '[A]fodillus anglice rames'. The list, which is disordered, full of errors and strangely deformed lemmata, ends at the letter 'r' and passes into medical receipts. Only very sparing use of it is made in the present study.

35 London, B.L. Harley 2378 ff. 113r–117r
A collection of medical treatises and receipts, the latter largely in English. After a *Quid pro quo* (ff.110v–113r) the list of synonyms begins with a red rubric *Hic incipit nomina herbarum* followed by the entry 'Ambrose .i. hynhole .i. wild sage' and concluding 'Explicit tractatus herbarum' (f.117r). The list is exclusively made up of *synonyma*. There are no descriptions and no use is made of *anglice/gallice*. The alphabetical sections are marked by a decorated initial in blue and red. The text is written consecutively across the page, headwords being distinguished simply by the size of their initial letter.
Lit. G. Henslow, *Medical Works of the Fourteenth Century together with a List of Plants recorded in Contemporary Writings, with their Identifications* (London, 1899), pp. 74–122.

36 London, University College lat. 11 ff. 335va–346vb
The MS is essentially a vast dictionary of medical terms (ff.1r–323v: Abhominacio – Vulva) which was completed in 1425 according to the explicit. The list

of *synonyma* begins 'Aaron barba Aron jarus pes vituli ȝekesters' and ends 'Explicit deus laudes 14 . . .' Each entry is preceded by a red paragraph mark. At the top of each column in red is the alphabetical letter of the section, the different sections being separated by blank spaces. The list is mainly of *synonyma*, but there are a few references to qualities (not degrees). Some additions have been made. Some columns have been badly affected by damp and discoloration.

Lit. D. K. Coveney, *A Descriptive Catalogue of Manuscripts in the Library of University College, London* (London, 1935), pp.10–11 & N. R. Ker, *Medieval Manuscripts in British Libraries* 1 (Oxford, 1969), p. 341.

37 London, B.L. Sloane 3217 ff. 1r–77v[15]

A text of the *Alphita* apparently, though frequently different from the text published by Mowat. Each entry begins with a red or blue initial and there are red paragraph marks and scrollwork throughout. The synonyms are given in Latin, French or English and, where they exist, descriptions are in English. These give degrees and qualities and are sometimes quite extensive, many of them being devoted to the animal and mineral world. The rest of the MS is made up of Middle English medical receipts.

38 Cambridge, Trinity College O. 8. 2 ff. 6ra–10va

A collection of medical texts. The list of *synonyma* begins 'Apium commune smalache'. There is no colour or decoration. Many of the entries contain Latin *synonyma* only. There are no descriptions or references to qualities and degrees. Like C1 above, to which it is affiliated, the list is rich in compound lemmata indicating varieties of a plant (i.e. through the use of epithets like *albus/rubeus*, *maior/minor*, *agrestis* etc.).

Lit. M. R. James, *A Descriptive Catalogue of the Manuscripts contained in the Library of Trinity College, Cambridge* 3 (Cambridge, 1902), pp. 388–89.

39 Cambridge, Trinity College O. 1. 13 ff. 28r–34r

A list, made up exclusively of synonyms, with the heading *Incipit exposicio sive interpretatio nominum herbarum secundum ordinem alphabeti et vocatur herbarium* and which begins 'Alleluya anglice wodsere [sic] vel kukkowmete'. There is no colour or decoration. Each alphabetical section begins with a capital letter in the ink of the text. There are some folios missing (e.g. between ff. 30 and 31) and many blank spaces are found in the lower halves of pages. Folio 29r/v contains diagrams and medical receipts and has nothing to do with the list of *synonyma*.

Lit. James, *Catalogue* . . . 3, pp.11–13

40 Cambridge, Trinity College O. 1. 13 ff. 37va–43rb

A volume of medical receipts. The list of *synonyma* begins 'Alleluya panis

[15] Cf. the similar text in MS London, B.L. Royal 12 D XII (s.xiv) ff.1ra–35vb which offers a few vernacular glosses e.g. f.19rb lupulus: hop-tre; f. 28va reginella: anglice medewort; f. 28vb resta bovis: caltrap; f. 34rb vicetorium: herbe Robert.

cuculi payn de cukelle wodesouwre scusteworte' and ends incomplete. In the beginning the headwords are English, with Latin and vernacular equivalents listed below them as in B4 above. But later there are entries which include glosses and descriptions which constitute the more conventional list of *synonyma*, which this text eventually becomes. There is no colour or decoration.
Lit. James, *Catalogue* . . . 3, pp.11–13.

41 Cambridge, Trinity College R. 14. 32 ff. 1v–65r
An English herbal with the heading *[H]ere begynnythe an herbale of namys and vertues of diverse herbys aftyr letterys of the a.b.c.* The heading is found at the top of f. 2r with 'Apium risus' as the first headword, but on f.1v there are entries for 'apium domesticum' and 'anetum'. The headwords are written larger and blacker than the text. There is no colour, the coloured initials for each entry having never been executed – only the guide letters remain. Single-line spaces separate the entries and some additional material has often been inserted in these spaces. Occasionally (e.g. f. 5r 'absinthium' and 'arnoglossa') the scribe forgot to emphasize the headwords and wrote them in the same manner as the text. A second hand is responsible for all the entries on f.5v and many elsewhere, whilst a third hand is seen on ff. 8r, 17v, 18r/v etc. From f. 42v onwards the space allotted for each entry has not been fully occupied, sometimes only a few lines being provided. Blank folios occur throughout the copy: ff.12r/v, 21v, 22v, 29r/v, 31v, 32r/v, 34v, 36v, 43r/v, 45v, 51v, 52v, 53r/v, 55v. Folio 19v contains 'virtutes croci' and f. 64r 'virtutes vervene'. The entries in the herbal are quite short, usually beginning with vernacular *synonyma*. The rest of the volume is made up of a paper MS containing various medical treatises in Latin and English.
Lit. James, *Catalogue* . . . 2, pp. 317–19.

42 London, B.L. Sloane 5 ff.13ra–57ra
An English herbal which, according to a colophon (f. 57ra) 'Johannes Lelamour scolemaister of Herforde est, þey [sic] he unworthy was, in the yer of our Lord anno mccclxxiii tournyd into ynglis'. There is no heading, but the text ends 'Explicit Macer'. The first entry is 'Ache is hote and (and) drye'. There are usually rubrics (the Latin plant name) in brown, underlined or boxed in red. The first word of each entry has a red initial letter and there are red paragraph marks in the body of the entry. There are 214 articles in all and various marginal annotations. The articles are not in alphabetical order. A folio is missing between ff. 19 and 20. There is an index on ff. 58vb–60vb.
Lit. Frisk, *A Middle English Translation of Macer Floridus* . . . p.16

43 London, B.L. Sloane 282 ff. 206v–210r
An abbreviated prose compendium of plant *synonyma* with a description of the medicinal properties and uses of each plant. There is a red rubric *Hic tractat Macer de virtutibus diversarum herbarum et primo de* Arthemisia anglice mogwort . . . The opening initial is gold, thereafter the initial letter of each headword is red or blue and the initial letter of each alphabetical section is treated similarly. There are no plant descriptions and the *synonyma* are introduced by

the unusual formula 'romani vocant eam . . ., anglici . . .' The short treatise is essentially a guide to the medicinal indications of each plant. The text is written in the same hand as the bulk of the MS which is a large collection of medical texts (see C10 above).

List of Incipits[16]

Aaron barba Aaron jarus pes vituli . . . (ThK 5) : B7, C5, C6, C8, C13
Abrotanum gallice averoigne . . . (ThK 10) : C30
Absinthium amarum . . . (ThK 11) : A3, B2, C1, C16, C18, C22, C23
Absinthium gallice aloyne anglice wermot . . . (ThK 12) : A1
Acacia .i. succus prunellarum agrestium . . . (ThK –) : A2
Acardella sowþystell . . . (ThK –) : C25
Ache is hote and drye . . . (ThK –) : C42
Aer idem quod divus . . . (ThK 69) : A10
Afodillus anglice rames . . . (ThK –) : C34
Alipiades .i. laureola herba catholica . . . (ThK 81) : C4
Alleluia anglice wodesour . . . (ThK 83) : B5, C10, C11, C24, C27, C39, C40
Alphita .i. farina ordei . . . (ThK 86) : C29, C37
Amarusca anglice maythe . . . (ThK 89) : A6, A8, A9, A12, B9, C21
Ambrose .i. hynhole .i. wild sage . . . (ThK –) : C35
Antemus .i. camamille . . . (ThK 107) : C14
Apium commune smalache . . . (ThK 113) : C38
Apium risus . . . (ThK 114) : C41
Artemisia anglice mugwort . . . (ThK 147) : B3, B6, C12, C28, C32
Artemisia domestica .i. febrifuga nostra . . . (ThK 147) : C7
Arthemisie tres species sunt . . . (ThK 148) : B1
Atriplex agrestis crisolecanna anglice meldys . . . (ThK –) : C3

In considering the value of the various testimonies offered by the MSS, albeit without establishing anything like a stemma, it is important to bear in mind the following manuscript affiliations:[17]
A3/C22; A9/A12; B1/C28; B2/C18; C5/C6 (very close identity); C5/C21; C6/C36; C10/C21/C24/C27; C21/C32.

[16] References in brackets are to L. Thorndike and P. Kibre, *A Catalogue of Incipits of Mediaeval Scientific Writings in Latin*, rev. & augm. ed. (London, 1963).
[17] These affiliations are based on the presence of common errors or the appearance of extremely rare lexical items.

The Principal Sources of Medieval Botany

Theophrastus

Theophrastus of Eresos (372/370 BC – 288/286BC) became head of the Peripatetic School in succession to Aristotle. His voluminous writings extend to over two hundred works and his *De historia plantarum* and *De causis plantarum* are amongst the best preserved. The botanical works were not translated into Latin until the 15th century, and in the 16th century many botanists returned to his work as an important authority. It was this aspect of Theophrastus's work which eclipsed all others in the period of the Renaissance.

Lit. C. B. Schmitt, 'Theophrastus' in P. O. Kristeller & F. E. Cranz (eds.), *Catalogus translationum et commentariorum: Mediaeval and Renaissance Latin Translations and Commentaries* II (Washington, D.C., 1971), pp. 239–322.
id., 'Theophrastus in the Middle Ages', *Viator* 2 (1971), 251–70.
A. Hort (ed. & transl.), *Theophrastus: Enquiry into Plants* (London, 1948).
J. Scarborough, 'Theophrastus on Herbals and Herbal Remedies', *Journal of the History of Biology* 11 (1978), 353–85.
G. Senn, *Die Pflanzenkunde des Theophrast von Eresos* ed. O. Gigon (Basel, 1956)

Dioscorides

After Hippocrates and Galen, Pedanius Dioscorides of Anazarbos (fl. 40–65 AD) is the most important ancient medical writer, and by far the most influential authority on *materia medica*, who was the fountainhead of the Western tradition of herbals. The *De materia medica*, composed c.64 AD and dealing (in Books 3 & 4) with around five hundred plants, has a most complex history and very little is known about its original and authentic form. It was very rapidly annotated, enlarged, modified and illustrated – as in the case of the famous Viennese codex Nationalbibl. med. gr. 1. Already by the second century a revisor had introduced alphabetical ordering of the materials. A set of glosses consisting of *synonyma* in various languages was early on affixed to the beginning of most chapters and they are indicated by 'Notha' in many editions. A Latin translation appeared as early as the sixth century, but this was superseded by a new translation, *Dioscorides alphabeticus*, c.1100, with stricter alphabetical arrangement. Shortly after 1300 this version was glossed by Petrus Padubanensis. Another Latin translation was apparently available to the Italian Rufinus in his *Liber de virtutibus herbarum* (c.1290) (q.v.). However, Dioscorides's greatest influence was exercised through piecemeal plundering of his work by almost all medical writers from 1000 to 1400. They particularly made use of a fifth-century text, the *Liber medicinae ex herbis femininis* which is based largely on the *De materia medica*. For 1500 years Dioscorides remained the undisputed master of *materia medica*.

Lit. J. M. Riddle, 'Dioscorides' in F. E. Cranz & P. O. Kristeller (eds.), *Catalogus*

translationum et commentariorum: Mediaeval and Renaissance Latin Translations and Commentaries IV (Washington, D.C., 1980), pp.1–143
id., Dioscorides on Pharmacy and Medicine (Austin, Texas, 1985)
J. Scarborough & V. Nutton, 'The Preface of Dioscorides' *De materia medica*: Introduction, Translation, Commentary', *Transactions and Studies of the College of Physicians of Philadelphia* 4 no. 3 (1982), 187–227.
C. E. Dubler, *La 'Materia médica' de Dioscórides: transmisión medieval y renacentista* 1–6 (Barcelona, 1953–59)
R. T. Gunther, *The Greek Herbal of Dioscorides: Illustrated by a Byzantine AD 512, Englished by John Goodyer AD 1655, Edited and First Printed by Robert T. Gunther* (New York, 1934; repr. 1959, 1968)
E Bonnet, 'Essai d'identification des plantes médicinales mentionnées par Dioscoride, d'après les peintures d'un manuscrit de la Bibliothèque Nationale de Paris ms. grec no. 2179', *Janus* 8 (1903), 167–77, 225–32.
O. Mazal, 'La forza delle erbe' [Il Dioscoride viennese codex Medicus graecus 1, Österreich. Nationalbibliothek] *Kos* 2, no.11 (1985), 17–32.
id., Pflanzen, Wurzeln, Säfte, Samen: Antike Heilkunst in Miniaturen des Wiener Dioskurides (Graz, 1981)
H. Gerstinger, *Der Wiener Dioskurides (Codex Vindobonensis Med. gr. 1)* (Graz, 1970)
J. M. Riddle, 'Pseudo-Dioscorides' *Ex herbis femininis* and Early Medieval Medical Botany', *Journal of the History of Biology* 14 (1981), 43–81.
J. Stannard, 'Dioscorides and Renaissance Materia Medica', *Analecta Medico-Historica* 1 (1966), 1–21.

Pliny the Elder

Gaius Plinius Secundus (c.23 – 79AD), the most important Roman author on natural science, wrote a vast *Historia Naturalis* in thirty-seven books, which maintained an unbroken tradition of popularity for 1500 years. Although often uncritical, Pliny was encyclopaedic in coverage and the work was constantly excerpted, the more so since its division of contents was so clear (animals and plants are treated in Books 20–32). In the third century appeared a pseudo-Plinian *Medicina Plinii* which, together with the original, became the source for the *Herbarius Apulei Platonici* or Pseudo-Apuleius, a collection of medicinal prescriptions rather than botanical descriptions, produced at some time in the fourth century, which supplies lists of synonyms and was then combined with *De herba vettonica liber* and the *Liber medicinae ex herbis femininis* to form a trilogy known simply as *Herbarium* and translated into Old English.

Lit. M. C. Martini, *Piante medicamentose e rituali magico-religiosi in Plinio* (Roma, 1977)
H. Zotter, *Antike Medizin: die medizinische Sammelhandschrift Cod. Vindobonensis 93 in lateinischer und deutscher Sprache* (Graz, 1980)
C. G. Nauert Jr., 'Caius Plinius Secundus' in F. E. Cranz & P. O. Kristeller (eds.), *Catalogus translationum et commentariorum: Mediaeval and Renaissance Latin Translations and Commentaries* IV (1980), pp. 297–422.

J. Stannard, 'Medicinal Plants and Folk Remedies in Pliny, *Historia Naturalis*', *History and Philosophy of the Life Sciences* 4 (1982), 3–23.
id., 'Pliny and Roman Botany', *Isis* 56 (1965), 420–25.
T. A. Sprague, 'Botanical Terms in Pliny's Natural History', *Kew Bulletin* 1933, 30–41.
H. J. de Vriend, *The Old English Herbarium and Medicina de Quadrupedibus*, EETS 286 (London, 1984)
F. W. T. Hunger, *The Herbal of Pseudo-Apuleius from the Ninth-Century Manuscript in the Abbey of Montecassino* (Leiden, 1935)
R. T. Gunther, *The Herbal of Apuleius Barbarus from the Early Twelfth-Century Manuscript formerly in the Abbey of Bury St. Edmunds (MS Bodley 130)* (Oxford, 1925)
H. Grape-Albers, *Spätantike Bilder aus der Welt des Arztes: medizinische Bilderhandschriften der Spätantike und ihre mittelalterliche Überlieferung* (Wiesbaden, 1977)

Isidore of Seville (c.560–636)

Isidore's *Etymologiae* remained the most celebrated encyclopaedia of the Middle Ages. The section on botany comprises Book 17, chapters 6–11. Here Isidore lists the names of plants, starting with the Latin name, then a Greek synonym, then an etymology, followed by a description and an enumeration of the plant's properties. His knowledge is notably less than Pliny's and is drawn entirely from books. He makes a considerable number of errors. His sources have not all been traced, but they certainly include an early Latin translation of Dioscorides, and do not include Pliny.

Lit. J. André (ed. & transl.), *Isidore de Séville: Étymologies Livre XVII* (Paris, 1981)
J. Fontaine, *Isidore de Séville et la culture classique dans l'Espagne wisigothique* 3 vols., 2nd. ed. (Paris, 1983)

Walahfrid Strabo

Walahfrid Strabo was Abbot of the Reichenau from 838-849 and in the 840s wrote a gardening poem entitled *Hortulus* in which he devotes stanzas to twenty-three plants. The plan of Walahfrid's garden, which he tended himself, can be compared with the *herbularius* in the famous monastic plan of St. Gallen, and the plants themselves with those (73!) prescribed in one of the laws of Charlemagne, the *Capitulare de villis* of c.800.

Lit. H.-D. Stoffler, *Der Hortulus des Walahfrid Strabo* (Sigmaringen, 1978)
R. Payne (transl.) & W. Blunt (comm.), *Hortulus: Walahfrid Strabo* (Pittsburgh, 1966) (includes facsimile).

V. Näf & M. Gabathuler (ed., transl., introd.), *Walahfrid Strabo: Hortulus, Vom Gartenbau* (St. Gallen, 1942)
Walahfrido Strabone, Hortulus tr. & cur. C. Roccaro (Palermo, 1979)

Macer

At some time in the eleventh century a poem of 2269 Latin hexameters entitled *De virtutibus herbarum* was composed by an anonymous author (who may have been Odo of Meung-sur-Loire) and later, perhaps c.1120, received the name Macer (in imitation of Aemilius Macer) and Floridus. The poem deals with approximately 77 herbs (depending on the MSS) in individual sections and draws material from Pliny, Dioscorides, *Medicina Plinii*, Pseudo-Apuleius etc. It is often close to, and it probably influenced, the *Flos medicinae Salernitanae*. It was frequently expanded, translated and excerpted and eventually became a set book at Montpellier. A modern critical edition is urgently needed.

Lit. Macer Floridus: De virtutibus herbarum. Traduzione e commento del dott. Roberto Trifogli (Roma, 1958)
L. Elaut, 'Parahistorisch Kommentar op een berucht farmakologisch Traktaat:de Macer Floridus', *Scientiarum Historia* 1 (1959), 149–59.
B.P. Flood Jr., 'The Medieval Herbal Tradition of Macer Floridus', *Pharmacy in History* 18 (1976), 62–66.
W. C. Crossgrove, 'Macer' in K. Ruh *et al.*, *Die Deutsche Literatur des Mittelalters: Verfasserlexikon* 5 (1985) cols. 1109–16.
id., 'Zur Erforschung des älteren deutschen Macer', *Sudhoffs Archiv* 63 (1979), 71–86.

'Circa Instans'

The work *De simplicibus medicinis*, usually known by its opening words as 'Circa instans', lists plants in alphabetical order and describes their properties and uses. It was composed by a member of the celebrated Salernitan medical family Platearius, either Johannes (d.1161) or Matthaeus, perhaps around the year 1150. Later it was used, along with Macer, for the *Regimen sanitatis Salerni*. It is the central work of a trilogy, which also includes the *Liber iste* and *Antidotarium Nicolai*, which remained the fundamental texts of medieval pharmacy for many centuries. There are over 200 MSS of the 'Circa instans', the earliest dating from the last decades of the twelfth century. The manuscript copies differ greatly in scope and arrangement, often reflecting the varying needs of the compilers and practitioners. In the early copies the text describes approximately 229 plants, adducing as authorities Dioscorides, Oribasius and various Salernitan masters. There are a number of medieval French translations.

Lit. G. Keil in K. Ruh *et al.*, *Die Deutsche Literatur des Mittelalters: Verfasserlexikon* 1 (1978) cols. 1282–85.

H. Wölfel, *Das Arzneidrogenbuch 'Circa instans' in einer Fassung des XIII. Jahrhunderts aus der Universitätsbibliothek Erlangen*, diss. Berlin, 1939.
A. Pfister, *De simplici medicina. Kräuterbuch-Handschrift aus dem letzten Viertel des 14. Jahrhunderts im Besitz der Basler Universitäts-Bibliothek* (Basel, 1960)
P. Dorveaux, *Le Livre des simples medecines* (Paris, 1913)

Hildegard of Bingen (1098–1179)

Hildegard founded a convent at Rupertsberg near Bingen in 1147 or 1148, becoming its first abbess. There she recorded her visions, wrote mystical works, homiletic and exegetical treatises and composed outstandingly beautiful religious songs. In addition she interested herself in natural history and medicine, writing *Liber simplicis medicinae* ('Physica') and *Liber compositae medicinae* ('Causae et Curae') between 1151 and 1158. They were originally united in a single work with the title *Liber subtilitatum diversarum naturarum creaturarum* of which a new edition is urgently needed. The first part (*Liber simplicis medicinae*) describes the curative properties of plants, trees, stones, metals and animals, following the teaching of the temperaments and humoral pathology. It covers 138 herbs. Little has so far been discovered about Hildegard's sources.

Lit. A. Forsius, 'Hildegard Bingeniläinen', *Hippokrates* 3 (1986), 44–51.
I. Müller, *Die pflanzlichen Heilmittel bei Hildegard von Bingen* (Salzburg, 1982)
A. Führkötter, *Hildegard von Bingen* (Salzburg, 1972)

Albertus Magnus (1193?–1280)

Albertus Magnus was an outstanding Dominican philosopher, theologian and scientist, teacher of Thomas Aquinas, who commented on the pseudo-Aristotelian *De plantis* (now attributed to Nicolaus of Damascus, 64BC – 4AD) in the Latin translation of Alfred of Sareshel. In Book 6 of his *De vegetabilibus*, composed c.1260, Albert provided an important series of plant descriptions (470 of them) which amount to a herbal and which themselves derive from herbals. Albertus describes the therapeutic uses of the plants and explains them, usually in humoralistic terms.

Lit. E. Meyer & C. Jessen (eds.), *Alberti Magni De Vegetabilibus libri VII* (Berlin, 1867; repr. 1982)
W. Fauser, *Die Werke des Albertus Magnus in ihrer handschriftlichen Überlieferung* (Aschendorff. 1982–)
J. Stannard, 'Identification of the Plants described by Albertus Magnus, *De Vegetabilibus* lib. VI', *Res publica litterarum* 2 (1979), 281–318.
id., 'Albertus Magnus and Medieval Herbalism' in *Albertus Magnus and the Sciences: Commemorative Essays* ed. J. A. Weisheipl (Toronto, 1980), pp. 355–77.
id., 'The Botany of St. Albert the Great' in G. Meyer & A. Zimmermann (eds.), *Albertus Magnus Doctor Universalis 1280/1980* (Mainz, 1980), pp. 345–372.

Rufinus

The Herbal of the Italian Rufinus completed soon after 1287 and preserved in a single, fourteenth-century codex (Ashburnham 189 in the Biblioteca Mediceo-Laurenziana) is remarkable for its original descriptions of plants wich seem to derive from personal observation and first-hand knowledge (Rufinus makes reference to the practices of herbalists at Bologna and Naples). At the same time its author conscientiously excerpts Dioscorides, Macer and *Circa Instans*, especially the latter, in a way which sheds light on their textual history. Remarkably, Rufinus does not seem to have known the work of Albertus Magnus.

Lit. The Herbal of Rufinus: Edited from the Unique Manuscript by Lynn Thorndike assisted by Francis S. Benjamin Jr. (Chicago, 1946)

Vincent of Beauvais

In his vast *Speculum maius* Vincent records more than 150 herbs, 170 domestic plants and 250 trees. His knowledge is derived entirely from books.

Tractatus de herbis

According to MS London, B.L. Egerton 747 (s.xiv) ff.1ra–106rb, a beautifully illustrated codex which stands between the *Circa Instans* (from which it borrows heavily) and the *Grant Herbier* (q.v.), this work is the compilation of Barthelemy Mini of Siena (s.xiii^{2}), a contemporary of Rufinus (q.v.). It has the prologue and general arrangement of the *Circa Instans*, but has been supplemented from other sources including Pseudo-Apuleius, Dioscorides and Macer. It is itself amplified in the Egerton MS by contemporary additions in the margins which are subsumed in the text of the other three known Latin MSS and in the French translation. For each simple the *Tractatus* offers an enumeration of synonyms and a brief description. The colophon of the Egerton MS runs: *Explicit tractatus herbarum, Dioscorides et Platone adque Galienus et Macronem translat(at)e manu et intellectu Bartholomei Mini de Senis in arte speciarie semper infusus. Deo gratias, Amen.*

Lit. G. Malandin, F. Avril, P. Leutaghi, *Platearius: le livre des simples medecines d'après le manuscrit français 12322 de la Bibliothèque Nationale de Paris* (Paris, 1986)

Bartholomaeus Anglicus

Bartholomaeus's *De proprietatibus rerum* was translated into English by John of Trevisa in 1398/9. Book 17 deals with plants and cites Dioscorides, Isidore, Platearius and others.

Lit. M. C. Seymour (gen. ed.), *On the properties of Things. John Trevisa's translation of Bartholomaeus Anglicus, De proprietatibus rerum*, 3 vols. (Oxford, 1975–

88) Cf. H. Boese (ed.), *Thomas Cantimpratensis: Liber de natura rerum* 7 (Berlin/ N.Y., 1973) Lib. XII.

Le Grant Herbier en François

Also known as the *Arbolayre*, this work was composed at the end of the 14th century and survives in c. 23 MSS from the following two centuries. It was first printed in 1488. Essentially it is a compilation based on the *Circa Instans* (q.v.) and *Tractatus de herbis* (q.v.) It was translated into English as *The Grete Herball* (q.v.) in 1526.

Lit. G. Camus, *L'Opera salernitana 'Circa Instans' ed il testo primitivo del 'Grant Herbier en Francoys'* (Modena, 1886)
Livre des simples medicines: codex Bruxellensis IV 1024. Texte et commentaires Carmélia Opsomer, préface Guy Beaujouan, 2 vols. (Antwerpen, 1980)
English translation of the above (which proposes some different plant identifications) by E. Roberts & W. T. Stearn (Antwerp, 1984)

The Grete Herball (1526)

This early printed work of 170 folio pages gives some idea of plant-name usage at the close of the Middle Ages. It contains over 500 English words for plants, fruits and spices. It is the first illustrated book on plants printed in England and contains 477 woodcuts. There are 505 chapters, mostly on plants, and arranged alphabetically (with some exceptions) according to the Latin names. The information recorded includes qualities and degrees, habitat, time of gathering, medicinal uses of the plant. The descriptions concentrate, as usual, on leaf shape and ignore floral structure. There is often an attempt to explain the motivation of the plant name. The work treats about 150 plants which might be termed British, offering approximately 500 English names and referring to some 300 species or genera.

Lit. M. Rydén *The English Plant Names in The Grete Herball (1526)*, Stockholm Studies in English 61 (Stockholm, 1984)

Synonymies

The need for synonymies – collections of *synonyma* – was felt early in the development of herbals. Early MSS of Dioscorides had been equiped with names from the first-century Alexandrian botanist and lexicographer Pamphilus (*Peri botanon*) and the Pseudo-Apuleius *Herbarium* lists *synonyma* (*nomina herbae*) at the end of each entry.[18] The great glossaries of the Carolingian period, particularly the *Liber glossarum* or 'Glossarium Ansileubi', contained plant names and *synonyma*

[18] In both works the *synonyma* are drawn from Pamphilus and include Gaulish, Etruscan, Dacian, Spanish, African and Syrian names.

amongst a significant number of medical glosses.[19] Curiously, there seems to be nothing ascribable to the twelfth century. As is the case with interlinear glossing in the vernacular, the provision of vernacular *synonyma* seems to have been the work of the thirteenth century. But there are monolingual Latin *synonyma* lists from an earlier period, as can be seen from an eleventh-century codex, MS London, B.L. Sloane 475 ff.143r–160r (inc. 'Anesus id est herba, folia eius coryandro simulant'). Single entries i.e. headword plus synonym characterize the twelfth-century Latin lists in MSS Cambridge, Trinity College B.1.39 ff.101v–103r (inc. 'Alleluia .i. panis cuculi') and O.3.42 ff. 99v–100r (inc. 'Amaracum .i. marmorana [corr. marjorana]'). The thirteenth century saw the rise of large glossaries incorporating the results of Arab medicine.[20] Soon after 1292 Simon Cordo of Genoa, physician to Pope Nicholas IV (1288–92), produced his *Synonyma medicinae* or *Clavis sanationis*, an attempt to bring together in a dictionary the medical and botanical terms found in earlier Greek, Latin and Arabic authors. To produce this work he travelled widely and became a patient observer of plants for thirty years. His valuable work was first printed in 1473 (the Venice edition of 1514 is inferior) and contained six thousand or more articles arranged alphabetically containing various observations, *synonyma*, and illustrative quotations. It was at once excerpted by Matthaeus Silvaticus of Mantua (d.1342) who began his *Pandectae medicinae* in 1297 and addressed it to Robert of Naples (1309–43). The first printed edition appeared in 1474. The 1488 edition contained about 720 entries, not always in strict alphabetical order. As well as travelling, Matthaeus possessed his own garden and imported seeds from Greece for it. His plant descriptions are often detailed and accurate. Mowat prints as an appendix to his *Alphita* a set of *synonyma* which he attributes to Peter of Abano (c.1250–c.1320), author of the *Conciliator*. Also edited by Mowat are the so-called *Synonyma Bartholomei*, which consist of c. 750 entries with largely single-word glosses. This list of *synonyma* is appended to the *Breviarium* in MS Oxford, Pembroke College 2 (s.xiv). The *Breviarium Bartholomei* is the work of John Mirfield (d.1407) who was a clerk at the Austin priory of St. Bartholomew, Smithfield, and subsequently at the hospital.[21] He probably came from Mirfield near Leeds in the West Riding of Yorkshire. His *Breviarium* is a rich source of knowledge of medicine in fourteenth-century England and deserves a full-scale study.[22]

19 Excerpted by J. L. Heiberg, *Glossae medicinales* (København, 1924) Cf. more generally L. de Man, *Middeleeuwse systematische glossaria* (Brussel, 1964), pp. 89–113.

20 See M. Steinschneider, 'Zur Literatur der Synonyma' in J. L. Pagel, *Die Chirurgie des Heinrich von Mondeville (Hermondaville)* (Berlin, 1892), pp. 582–95. On twelfth-century interest in Arab botany see J. H. Harvey, 'Gardening Books and Plant Lists of Moorish Spain', *Garden History* III, 2 (1975), 10–21 and *id.*, 'Turkey as a Source of Garden Plants', *ibid.* IV, 3 (1976), 21–42. See also L. C. MacKinney, 'Medieval Medical Dictionaries and Glossaries', in J. L. Cate & E. N. Anderson (eds.), *Medieval and Historiographical Essays in Honor of James Westfall Thompson* (Chicago, 1938) [pp. 240–68] pp. 261ff.

21 See P. H.-S. Hartley & H. R. Aldridge, *Johannes de Mirfield of St. Bartholomew's, Smithfield: His Life and Works* (Cambridge, 1936).

22 The only recent publication is F. Getz, 'John Mirfield and the *Breviarium Bartholomei*: The Medical Writings of a Clerk at St Bartholomew's Hospital in the later Fourteenth Century', *The Society for the Social History of Medicine, Bulletin* 37 (1985), 24–26. See also J.

By the thirteenth century the interest in lexical matters becomes prominent in many collections of *synonyma*. Commenting on Byzantine botanical lexica of this period Stannard declares 'Lexicographical considerations have virtually displaced the botanical data. Most of the glosses are bare synonyms and only infrequently do they contain descriptive data pertaining to the plant name of the lemma. Moreover, the preoccupation with lexical matters is carried to the point of recording synonyms from several languages . . .'[23]

Vernacular plant lists have not been adequately studied in a European context. It is clear that they survive in a wide range of languages. In MS London, B.L. Arundel 295 (s.xiv) ff.121ra–130vb there is a Latin-German plant vocabulary (inc. 'Aaron .i. jarus abrotanda scutewrtz') in which the Latin headwords are glossed with Latin and vernacular equivalents. The list is then followed (ff.131ra–133vb) by a list of German words with abbreviated indications of where they will be found in the Latin-German section e.g. 'borncrisse: na[sturcium], se[nacion]'.[24] There is a Latin-Dutch vocabulary in MS Sloane 345 (s.xv) ff.79r-83r.[25] What is really needed is a handlist of vernacular *synonyma* collections.

The Plant Names in the Synonyma Herbarum

The absence hitherto of any printed corpus of Middle English plant names[26] has inevitably led to a number of misapprehensions concerning the development of English plant nomenclature in more recent times. Ryd[ac]en, for example, lists 124 English plant names (many of which are familiar names preceded by the epithets little-, great-, wild-) which he considers to be first attested in *The Grete Herball* of 1526.[27] The following 15 words from his list are already found in our medieval *synonyma* lists: bear's-foot, bistort, caseweed, croyt marine, cuckoo's- bread, fume-of-the-earth, hare's-lettuce, hog's-meat, king's-crown, knotgrass, polytrich, remcope, Solomon's seal, scrophulary, tintymall. Rydén also considers that Gerarde provides the first records of 182 native British species,[28] but the fact remains that the debt of the sixteenth-century English herbalists to their medieval

B. Colton, 2nd., *John of Mirfield (d.1407) Surgery: a translation of his' Breviarium Bartholomei', part IX* (New York, 1969)

23 J. Stannard, 'Byzantine Botanical Lexicography', *Episteme* 5 (1971) [168–87] 179–80.

24 See R. Priebsch, *Deutsche Handschriften in England* 2 (Erlangen, 1901), no. 72, pp. 49–50.

25 See *ibid.*, p. 23. Cf. the Dutch herbal in L. J. Vandewiele, *De 'Liber magistri Avicenne' en de 'Herbarijs'*, pt. 2 (Brussel, 1965) & *id.* & W. Braekman, 'Een latijns-mnl. plantenglossarium uit het midden van de 14e eeuw', *Scientiarum Historia* 10 (1968), 115–44.

26 J. Earle's *English Plant Names from the Tenth to the Fifteenth Century* (Oxford, 1880) merely reproduces previously printed materials and includes only four modest lists of Middle English plant names, three of which date from the fifteenth century.

27 See M. Rydén, *The English Plant Names in the Grete Herball (1526)* (Stockholm, 1984), pp. 35ff. Rydén comments 'The percentage of ME names in GH [= Grete Herball] is probably higher than 55% [= the figure he has computed] since some of the names first attested in GH, though not evidenced (so far) in ME records, are presumably of ME heritage'.

28 *ibid.*, p.11 n. 7.

predecessors has never been investigated. A glance at Grigson's *Dictionary of English Plant-Names* (1974) reveals that many of his datings are hopelessly late, but this is because the OED is deficient in the medieval period. Thus, the following names from our *synonyma* are first recorded in the OED from the 16th century: all-heal, argentil, argentine, asarabacca, bearwort, bear's-foot, bird's-nest, blue-bottle, brank-ursine, camel's-hay, camel's-meat, case-weed, churchwort, cock's-comb, cock's-foot, colt's-foot, couchgrass, crete marine, crosswort, crow's-toes, cuckoo-flower, dewberry, epatic, erynge, eupatory, exan, eyebright, filipendula, fishmint, forebitten, gallwort, gandergoose, glaunde, goat's-beard, goose-chite (-clite in my MSS), hare's-ear, hare's-lettuce, hare's-thistle, hazelwort, horse-chire, horse-tail, hurt, lamb's-tongue, lark's-foot, larkspur, lion's-tooth, lodewort, mad-wort, French mallow, masterwort, milkwort, moonwort, man's motherwort, nail-wort, pepperwort, peterwort, pick-purse, polytrich, potwort, psylly, ramp, red-knees, saint Mary's seed, sanagre (savager in Gerarde), sea-rush, sea-willow, sea-wormwood, setterwort, shamrock, shepherd's-pouch, Solomon's seal, spignel, spurgewort, squinant, standelgrass, standelwort, stanchblood, steadfast, stone-break, stonewort, stubwort, swordgrass, toothwort, tree-of-paradise, turmeric, wake-robin, water-betony, water-parsley, water-rose, white wort, wolf's-fist. This total of 89 words seems small when compared with the nearly 500 items which are not recorded in the OED at all!

agnes cast, aldwort, alne, alomoth, alrewort, amarok, amys, anagal, andrage, anke-wort, appelwode, ass-thistle, asses' lettuce, banculwort, bear's-cheek, bell-weed, bellwort, beowort, bird's-tongue, bisseplese, biting-grass, bitterwode, bladwyne, bladyn, blowort, blue buckthorn, blue weed, bodwine, bolle-weed, bonrefair, bose-leye, bothen (blue-, gold-), botuner, sea-briar, broad-leek, brook-cress, brook-leaves, brook-lily, brook-meadwort, brookwort, brown sourwort, browort[?], bruer, brusk, bryndhale, bud-thistle, bun-thistle, buter besome, cabbagewort, ca-meos, campwort, cantabel, cardiac, carlowre, carter's-sage, casberry, cassia-fistre, cat-pipes, cat's-joy, centonicle, centoys, centregalle, cetelworth, chaunet, cheme-rek, cherl, chirsnedle, cilsper, citrine, clifweed (-wort), clodwort, cloter, cock's-spur, cockwort, codweed, colencrop, colt's-tail, copere, copwort, cordial, corneþuohe, count-mint, countwort, cow-eye, cow-meat, cow-weed, cow-wort, crake-foot, crake-leek, crawanbeam, crebyld, Cretan (pety-, water-), Cretan of the sea, cristiane, cripplewort, crisp-hock, crispine, crois(e, croiser, crowgold, crow-leaf, crow-rattle (red-, white-), crow-wort, cuckoo-bell, cuckoo-sour, cuckoo-wort, cuculeyne, curungen, culbertull, cywort, deadman-flower, devel's needle, diswort, dock-flower, dog's-ballocks, dolhrune, donnhove, dovewort, dowechs, doworte, dumb-nettle, earth-needle, earthwind, earwort, eble, eggid-gras, eglamour, ellern-wort, elmewort, emwort, ende-gras, ende-mete, ende-schede, enne-leac, eoforþrote, ersope (ursope), erswort, esule, evelþroten, everwort, everzwort, eye-wort, faitor's-weed, faitor's-wort, farthing-grass, favede, feterwort, field-bladder, field-more, fiend's-bite, fill, flebat[?], floatwort, flower-of-paradise, foal's-leek, foal's-tail, foal's-tongue, foam-dock, foseyn[?], foul-bean, fowl's-spur, fowl's-wise, fox-eye, fox's-foot, fox's-tongue, fraghan, fume-of-the-earth, galli-trik, garclive, saint Mary garlic, gazer, genicle, gersk, giggeflour, glewort, glove-wort, goat's-tongue, gold-cress, goldwort, gostel-tre, gourd-grass, gowk's-pint, gowk's-sour, grace-dieu, graciose, gretelwort, guþecorn(e, halsnote, hardwort, hare-teasel, hare's-glove, hare's-grass, hare's-mint, hare's-wort, hart's-foot, hart's-grass, hart's-puliol, harvest-plant, hawenehydele, hayleworte, healbrede, hedge-cleavers, helow, water hellebore, hen-cress, hen-penny, hen-wort, herb(e copi(e, herbe coupe, herb(e croys(e, herb(e Paul, herb pernell, herbe Salamon, her-mole, herswet, hethnote, hewen, hind-ladder, hog's-meat, hog's-pint, holkemeny, holleryd, holow-berys, holy-gold, holy-mallow, holy-whey, holywort, honey-suckle-of-the-hedge, horn-whistle, hors-fartis, horse-ear, horse-grass, horse-ivy, horse-tooth, hote-note, houderugresse[?], hound-flowers, houndhore[?], hound's-dill, hound's-head, hound's-heart, hound's-nettle, hound's-rose, hound's-sene-mayn(e, hound's-tind, hound's-tooth, house-lily, hundes-quelk, hundred-hole, hyerwort, ilhaw, kenningwort, kerwort, king's-crown, king's-wort, great knee-wort, knotting-grass, kympilmore, lady's-grass, lady's-ladder, lamb's-foot, lamb's-heart, langwort, lathwort, lesior, leverwort, lithwort, lithewort, littlewort, lobbe, locher[?], loudwort, louse-herb, lovewort, maidenwort, field-mallow, holy-mallow, marceure, marchwort, mareblowre[?], masewort, maturell, maythewort, mead-flower, mead foal's-foot, meald[?], mecalayn, meek, melizane, melose[?], memithe, menphie, meselwort, micklewort, middewort, mild-maid, milterdene (multerdene), mint Rome, mixberry, mixplant, mock, modermerche, molde, moon-herb, morsulk, morun, ragged motherwort, woman's motherwort, mouse-fitch, mouse-nest, mouse-pipe, much-dock, mulleinwort, neat's-tongue, needlewort, neppe ryall, nesh rushes, netebroun, nosegrass, okelescons, oumener, Our Lady's milk, ovensharp (-dock), owl-pipe, pakelwort, peacock's-tail, pentifelon, perdocal, pigell, pigment, pigmentary, pig's-bread, pirole, plaggell[?], pobel, podyll, polica-ri(e, portyn[?], poseire, pot-thistle, powk-cheese, puck's-bite, puttock's foot, qu-aterspys, radiol, raggedwort, raicele[?], rammes-lek, rancl wort, ranculwort, rap-coyle, rascal, raven's-leek, rayge, recoupe, rest-beof, reveray, roal, rodelet, romgresse, rose-grass, rote scabious, runcele, sacherey, saint John's herb, saint Mary's berry, saint Mary's flower, saint Mary's gold, saint Mary's ladder, saint Mary's rud, saint Mary's wort, sambuk, sanagre, sanigre, sanylle[?], sauuger, say-wort, scale-grass, scaverole, schilwort, scrophulary, sea-dock, sea-fike, sea-hock, sea-withy, seed-mint, semlestre[?], sent-chef, seofonleafe, shalwort, shepes-wort[?], sickleweed, siserioys[?], siwariswort, sleepingwort, slevewort[?], slewort, slicewort, slite, smallwort, smerdesworthe, snailwort (great-, little-), snap-grass, snapwort[?], soft, softwort, Solomon's throne, sparrowhawk's-foot, sparrow's-foot, speringwort, spewingwort, stalestole, stammeringwort, stanchwort, standel-gourd, stand-gourd, sterching-grass, stevenwort, stichegresse, sticklewort, stick-pile, suche-gresse, sulfurat, swine's-neep, swine's-tail, swordwort, symon(t), simonwort, tayweed, thringes, tie-herse, timbre, toad-berry, toad's-fennel, trap, tree-ivy, tronary (great-), tuteceleyne[?], uenwort, unfortredde, ungylwort, unset wort, ventolayne[?], water-bardan, water-clete, water-fike, water-horsetooth, water-march, water-rams, water-spearwort, wealmore, weasel-tongue, weyhore, white herb, wild hop(s), wild molece, wodewort, wolf-dock, wolf's-bollock, wolf's-comb, wolf's-cul, wolf's-tail, wolf's-teasel, wolf's-tooth, woodbane, wood-beard, woodbell, woodberry, woodbrake, woodbroom, woodclete, woodclot, wood-crowfoot, wood-fennel, wood-hache, woodhove, wood-ivy, woodkey, wood-lettuce, woodpipe, woodrattle, woodswythe, woodthung, wood-wort, wood-yarrow, wringingwort, wropewort, yardhove, yeke, yellow bon, yellow cray, yetewort.

Even allowing for the eventual removal of 5% of these names as scribal fantasies or editorial errors,[29] it is still a remarkable list which illustrates yet again how premature the elaboration of new dictionaries is when there is still so much to be gleaned from unread texts. Similarly, the writing of historical accounts is always subject to caution since the dating of lexical items is constantly being revised. The following examples from our *synonyma* will serve as illustrations. The date in brackets is that of the first recorded example given in the OED: arbust (1658), bearbind (1732), breakstone (1688), chickenwort (1765), Christ's-eye (no example provided), earth-apple (= sowbread, 1601), flaxwort (1846), goose-tongue (1738), gowk's-meat (1879), hog's-grass (1880), hungerweed (1793), kemps (1825), knotwort (1845), lentildew (1800), monk's-head (1682), needle (1793), nettlewort (1834), paddock-pipe (1673), ragweed (1658), renning-wort (rennetwort) (1727), scalewort (1849), shepherd's-staff (1760), springwort (1889), stanch (1726), stanchblood (1601), taperwort (1601), teazelwort (1846), water-bean (1846), (?)wartgrass (1880), water-hemp (1796), water-nut (1617), water-weed (1842), windgrass (1847), yellow-berry (1712).

A proper analysis of the plant names of medieval England must await the assembly of a fuller corpus of materials than is provided here. Nevertheless a number of preliminary observations may be made on the evidence of the *synonyma* lists. Of just over 1800 vernacular names approximately 400 are French, that is to say of unambiguously French form or else preceded by *gallice* in the MSS.[30] Whilst most of them are familiar from continental French there remains a number which are problematic e.g. barberole, barbote, bere, blabeus, bynde de urse, chilizene, cicuse(?), cifer, coper(e, elfione, gauge, gaure, gazer, haynere, her de fer(?), hermodie, melie/med(i)e, meraloigne/merlynge (read amer aloigne?), osatille, patracion, polion, rasteine, reblet, rubeseym, sacherey, scard, sclavelor.

Whilst the number of names given to a single plant does not achieve the levels recorded by Grigson or Britten and Holland (the latter note, for example, that Caltha palustris L. has 50 different names) for popular and dialect names in the modern period, there is still a corpus of commonly mentioned plants known by a variety of names.[31] The following list records the number of Latin names ascribed to a selection of plants from the *synonyma* lists: Adiantum capillus-Veneris L. (14), Althaea officinalis L. (16), Anthriscus sylvestris (L.) Hoffm. (15), Arctium lappa L. (22), Artemisia vulgaris L. (17), Asperula odorata L. (14), Bryonia dioica Jacq. (16), Calendula officinalis (16), Centaurium erythraea Rafn (13), Cichorium intybus L. (13), Dipsacus fullonum L. (13), Galium aparine L. (13), Hyoscyamus

[29] Palaeographical errors perpetuated in the MSS as living names include salmenta for saliunca (< saliūca); sanamenda for sanamunda (< sanamūda); grandin for gratia dei (< gr̄a dī); honstonge for lion's-tooth (< lionstoþe); wild- frequently misread as wud- or child-. Cf. also red clote for red doce; brokwort, broyswort, brothwort for brotherwort (broþ'wort).

[30] All the figures quoted here are of course approximate because of the problems of making sure identifications.

[31] Rydén, *op. cit.*, p. 49 claims that in the *Grete Herball* 59% of plants have one vernacular name, 23% two, and 18% three or more (to a maximum of five).

niger L. (13), Lolium temulentum L. (14), Lonicera periclymenum L. (20), Mercurialis perennis L. (16), Pastinaca sativa L. (13), Polygonum aviculare L. (16), Primula veris L. (13), Sempervivum tectorum L. (19), Solanum nigrum L. (14), Sonchus oleraceus L. (16), Verbascum thapsus L. (15). By contrast, few plants have more than five vernacular names, the most striking cases being Acanthus mollis L. (8 as against 3 Latin names), Anthemis cotula L. (10), Arum maculatum L. (at least 16), Erodium moschatum (L.) L'Herit. (11 as against 3 Latin names), Calendula officinalis L. (10), Capsella bursa-pastoris (L.) Medic. (12). Until a much fuller corpus of material is available it would be misleading to elaborate such lists and premature to assess their implications.[32]

There is an intractable group of polysemous names which belong to what Earle called 'a large circumambient zone of what may be called volatile names'.[33] In many cases a sure botanical referent cannot be attached to the name because of the complications of regional (synchronic) and chronological (diachronic) variation or the intrinsic generality of the name itself. Such names (with the number of Latin equivalents in brackets) are: bishopswort (17), bitterwort (6), bonewort (9), feverfew (9), foxglove (13), herb ive (12), hollyhock (11), motherwort (12), mugwort (22), orvale (6), scabwort (9), smearwort (12), wild neep (18), woodruff (19).

Another category of names consists of straight translations or minimal modifications of Latin names. They resemble the book-names provided in many nineteenth-century botanical works and are unlikely to have been in wide use: agnes cast, amyas, anagal, andrage, bractes, brank-ursine, centregalle, cretan of the sea, epatic, erynge, eupatory, filipendula, flos campi, matricary, memithe, perdical, policari(e, polytrich, portulac, scrophulary, tithymal. Less learned forms are also sometimes straight translations of a classical name e.g. asses' thistle (onopordon).

A fundamental problem is created by the fact that the botanical lexis of medieval Latin is no better established than that of Middle English and chronological data are so rare that we cannot usually determine in which direction influence has been exercised. The following Latin names seem to be fairly clearly modelled on the vernacular: achia, aragia, falgeria, faverola, frasearia, gletinaria, gaisdo. But how are we to order chronologically the many cases like herba auricularis/earwort; barba silvana/woodbeard; allium corvorum/crow-garlic; dens equi/horsetooth; auris equina/horse-ear etc.? Are we to postulate a *lingua pulli* behind the name 'foal's-tongue'? Leaving aside the survivals from the Old English botanical lexis,[34] it is not easy to establish the independent creation of vernacular names. One recurring phenomenon is the multiplication of variant forms of a common name: thus branca ursina gives rise to bear's-cheek, bear's-foot, and bearwort; orobus (auricula muris) is rendered by mouse-ear, mouse-fitch, mouse-pea and mouse-pipe; satirion is glossed by standelgrass, standelguss, standel-gourd, standel-welk and standelwort.

The motivation of plant nomenclature is a fascinating, but necessarily specula-

32 It is particularly difficult to evaluate the correlation of vernacular names and Latin names. For example, the word *clote* is used to gloss 26 Latin names (at least four of these cases involve errors), *henbane* accompanies 14 Latin names and *gladden* 15.

33 *op. cit.*, p. lxvi.

34 e.g. crawanbeam, hawenehydele, seafonleafe.

tive, subject which involves the study of folklore, popular religion and a host of other fields.[35] It is not in the least surprising that a large number of vernacular plant names derive from a fancied resemblance between the shape of some part of a plant – leaf, roots, fruit etc. – and an anatomical feature of a common animal or bird (cock, crow, dog, foal, fox, hare, hart, lamb etc.). The most productive suffixes are -foot, -tail, -tongue, and -ear. The most commonly utilized animals and birds are hare (11), horse (12), crow (11), hound (14), cuckoo (8), and wolf (10). Once again these figures do not achieve the levels recorded by Grigson in his lists of the popular and local names of plants,[36] where the corresponding figures are 10 (hare), 35 (horse), 23 (crow), 1 (hound; but 56 involving dog!), 35 (cuckoo) 0 (wolf). There are further striking contrasts. Grigson records 47 names in cow- and 61 in Lady's-. The intriguing question of the continuity or discontinuity of medieval and popular names is one that can scarcely be tackled in the present state of plant-name research, but will certinly be an important topic for future researchers working with more extensive materials.

Another significant derivational source is represented by medical indications, our *synonyma* lists being largely compiled as a key to the ingredients of medical receipts. Thus a series of plants are named after the parts or conditions which they are considered to heal or the effects they are said to produce:[37] arsesmart, bloodwort, bonewort, bruisewort, cancerwort, eyebright, gallwort, louse-herb, lousewort, lungwort, nailwort, nose-grass, ranculwort, sleepwort, spewingwort, stammeringwort, tetterwort. This custom is related to the so-called doctrine of signatures or phytognomy, the view that plants exhibit some external sign of their properties and the use to which they may be put.[38] This theory, which can be traced back in some form to sympathetic magic and the writings of Pliny the Elder, was promoted in the sixteenth century by the egregious Theophrastus von Hohenheim (1493–1541) who called himself Paracelsus. Thus red-leaved plants e.g. red cabbage reveal their utility for staunching blood or dressing wounds; plants which secrete a milky latex are indicated for nursing mothers; plants with snake-like roots assist in the treatment of snake-bites; the knotty tubers of the genus Scrophularia suggest their use for treating scrofulous glands; the scaly pappus of Scabiosa offers a treatment for leprosy; the hard, stone-like seeds of the Gromwell reveal it to be an effective lithagogue e.g. in the treatment of gravel.

These observations, intentionally slight, give some idea of the fascinating inter-

35 Cf. G. M. Awbery, 'Plant Names of Religious Origin in Welsh Oral Tradition' in R. Vickery (ed.), *Plant-Lore Studies*, Folklore Society, Mistletoe Series 18 (London, 1984), pp.1ff. and B. Morris, 'The Pragmatics of Folk Classifications', *ibid.*, pp.120–38. Cf. also A. Wölfel, *Pflanzen in der Volksmedizin. Die volksmedizinischen Studien von Heinrich Marzell*, diss. Erlangen – Nürnberg, 1983.

36 See G. Grigson, *An Englishman's Flora* (London, 1955).

37 A classification of plants according to their properties as haemagogues, hydragogues, phlegmagogues, purgatives, diuretics, styptics, emetics etc. is included in the so-called 'Salernitan Tables', as found, for example, in MS London, B.L. Sloane 3550 (c.1300) ff. 77r–84r. Plants are also listed according to their qualities and degrees.

38 See D. Evans, 'The Doctrine of Signatures as the Explanation of some puzzling Names and Uses of Plants' in R. Vickery, *op. cit.*, pp. 66–74. Cf. S. Mahdihassan, 'A Pahlavi Antecedent of the Doctrine of Signatures', *Hamdard Medicus* 24, iii–iv (1981), 31–5.

pretative problems which await the patient labourer who will first rescue the medieval plant names from the obscure documents in which so many of them remain buried.

BIBLIOGRAPHIES

DICTIONARIES OF PLANT-NAMES[39]

Anell, I., *Vad betyder växtens latinska namn* (Stockholm, 1976)
Berger, F., *Synonyma-Lexikon der Heil- und Nutzpflanzen* (Wien, 1954–55)
Bierbaumer, P., *Der botanische Wortschatz des Altenglischen* 3 vols., Grazer Beiträge zur englischen Philologie 1–3 (Bern/Frankf. am M., 1975–79)
Britten, J. & Holland, R., *A Dictionary of English Plant-Names* (London, 1886; repr. 1965)
Cameron, J., *The Gaelic Names of Plants (Scottish, Irish and Manx)*, new & rev. ed. (Glasgow, 1900)
Earle, J., *English Plant Names from the Tenth to the Fifteenth Century* (Oxford, 1880)
Ekwall, E., *The Concise Oxford Dictionary of English Place Names* (Oxford, 1936)
Fischer, H., *Mittelalterliche Pflanzenkunde* (München, 1929; repr. 1967)
Fisher, R., *The English Names of our commonest Wild Flowers* 1–2 (Arbroath, 1932–33)
Genaust, H., *Etymologisches Wörterbuch der botanischen Pflanzennamen* (Basel/Stuttgart, 1976)
Grassman, H., *Deutsche Pflanzennamen* (Stettin, 1870)
Grigson, G., *A Dictionary of English Plant-Names (and some Products of Plants)* (London, 1974)
Henslow, G., *Medical Works of the Fourteenth Century, together with a List of Plants recorded in contemporary Writings, with their Identifications* (London, 1899)
Howes, F. N., *A Dictionary of Useful and Everyday Plants and their Common Names* (Cambridge, 1974; repr. with corr., 1975)
Kadans, J.M., *Encyclopedia of Medicinal Herbs* (Wellingborough, 1979)
Keogh, J., *Botanalogia universalis hibernica* (Corke, 1735)
Loewe, R., *Germanische Pflanzennamen* (Heidelberg, 1913)
Marzell, H., *Wörterbuch der deutschen Pflanzennamen* 5 vols (Leipzig etc. 1942–79)
Mills, S., *The Dictionary of Modern Herbalism* (Wellingborough, 1985)
Moldenke, A. L. & H. N., *Plants of the Bible* (New York, 1957)
Plowden, C. Chicheley, *A Manual of Plant Names*, 3rd ed. (London, 1972)

[39] For the dictionaries of André and Carnoy see above p. xvi.

Prior, R. C. A., *On the Popular Names of British Plants* (London, 1863; 1870²; 1877³)
Pritzel, G. & Jessen C., *Die deutschen Volksnamen der Pflanzen* (Hannover, 1882)
Rayner, J. F., *A Standard Catalogue of English Names of our Wild Flowers* (London, 1927)
Reier, H., *Die altdeutschen Heilpflanzen, ihre Namen und Anwendungen in den literarischen Überlieferungen des 8.–14. Jahrhunderts*, 3 vols. (Kiel, 1982–83)
Rendall, V., *Wild Flowers in Literature* (London, 1934)
Rydén, M., *Shakespearean Plant Names: Identifications and Interpretation* (Stockholm, 1978)
id., The English Plant Names in The Grete Herball (1526): A Contribution to the Historical Study of English Plant-Name Usage (Stockholm, 1984)
Smith A. W. & Stearn, W. T., *A Gardener's Dictionary of Plant Names* (London, 1972)
Strömberg, R., 'Griechische Pflanzennamen', *Göteborgs Högsskolas Årsskrift* 46 (1940), 1–190
Thompson, R. C., *A Dictionary of Assyrian Botany* (London, 1948)
Threlkeld, C., *Synopsis stirpium hibernicarum* (Dublin, 1727)
Voss, E. G. *et al.*, *International Code of Botanical Nomenclature* (Utrecht, 1983)
van Wijk, H. L. G., *A Dictionary of Plant Names*, 2 vols. (The Hague, 1911–16)
Willis, J. C., *Dictionary of Flowering Plants and Ferns* (Cambridge, 1951)
Zohary, M., *Plants of the Bible* (Cambridge, 1982)

SOME MEDIEVAL PLANT GLOSSARIES (EDITIONS)

Floyer, J. K. & Hamilton, S. G., *Catalogue of Manuscripts preserved in the Chapter Library of Worcester Cathedral* (Oxford, 1906), pp.184–5 (= MS F.157 f.2ra/va, a list which appears to be full of errors)
Förster, M., 'Mittelenglische Pflanzenglossen (um 1300)', *Anglia* 42 (1918), 156–62.
von Lindheim, B., *Das Durhamer Pflanzenglossar*, Beiträge zur englischen Philologie 35 (Bochum, 1941; repr. 1967)
Mowat, J. L. G., *Sinonoma Bartholomei*, Anecdota Oxoniensia, Med. & Mod. Ser. I, 1 (Oxford, 1882)
id., Alphita. A Medico-Botanical Glossary, Anecdota Oxoniensia, Med. & Mod. Ser. I, 2 (Oxford, 1887)
Norwood, F. E. S., 'The Fourteenth-century Medico-Botanical Glossary of Raimon de Castelnou', *Proceedings of the Leeds Philosophical and Literary Society* 15 (1972/73), pp.135–223.
Stannard, J., 'A Fifteenth-Century Botanical Glossary (Huntington Library MS HM 64)', *Isis* 55 (1964), 353–67.
Stokes, Wh., 'A List of Welsh Plantnames from a fourteenth-century manuscript in the British Museum (Mus. Brit. Addl. 14,912)', *Archiv für celtische Lexikographie* 1 (1900), 37–49.
id., 'Three Irish Medical Glossaries', *ibid.*, 325–47.

id., 'On the Materia Medica of the Mediaeval Irish', *Revue celtique* 9 (1888), 224–44.

Stracke, J. R., *The Laud Herbal Glossary* (Amsterdam, 1974)

Vandewiele, L. J. & Braekman, W., 'Een latijns-mnl. planten-glossarium uit het midden van de 14e eeuw', *Scientiarum Historia* 10 (1968), 115–44.

Wilson, E., 'An Unpublished Alliterative Poem on Plant-Names from Lincoln College, Oxford, MS Lat. 129 (E)', *Notes & Queries* 224 (1979), 504–08.

Wright, Th. & Wülcker, R.P., *Anglo-Saxon and Old English Vocabularies* 2nd ed. vol.1 (London, 1884) cols.554–59 (= MS Harley 978)

HERBALS (STUDIES)

Anderson, F. J., *An Illustrated History of the Herbals* (New York, 1977)

Arber, A., *Herbals, their Origin and Evolution – a Chapter in the History of Botany 1470–1670* (Cambridge, 1912; 1953²; 1987³)

Blunt, W., *The Art of Botanical Illustration* (London, 1950)

id., & Raphael, S., *The Illustrated Herbal* (London, 1979)

Callery, R. G., 'Ancients, Moderns and Reliable Friends: Stated Sources of some English Herbals', *Journal of the Society for the Bibliography of Natural History* 8 (1978), 435–44.

Dawson, W. R., 'The Origin of the Herbal', *Aegyptus* 10 (1929), 47–57.

Grieve, M., *A Modern Herbal* (Harmondsworth, 1980)

Heilmann, K. E., *Kräuterbücher in Bild und Geschichte* (München-Allach, 1966; repr. 1980)

Henrey, B., *British Botanical and Horticultural Literature before 1800*, 3 vols. (London, 1975)

Kuźnicka, B., 'The Earliest Printed Herbals and Evolution of Pharmacy', *Organon* 16/17 (1980/81), 255–66.

Lawrence, G. H. M., 'Herbals: their history and significance' in *History of Botany: Two Papers presented at a Symposium held at the William Andrews Clark Memorial Library, December 7, 1963* (Los Angeles/Pittsburgh, 1965)

Nissen, C. *Die botanische Buchillustration. Ihre Geschichte und Bibliographie* 2 vols. (Stuttgart, 1951); 2nd. ed. and Supplement (Stuttgart, 1966)

id., *Herbals of Five Centuries: A Contribution to Medical History and Bibliography* (Zurich, 1958)

Rohde, E. S., *The Old English Herbals* (London, 1922)

Singer, C., 'The Herbal in Antiquity and its Transmission to later Ages', *Journal of Hellenic Studies* 47 (1927), 1–52.

Stannard, J., 'The Graeco-Roman Background of the Renaissance Herbal', *Organon* 4 (1967), 141–45.

id., 'The Herbal as a Medical Document', *Bulletin of the History of Medicine* 43 (1969), 212–20.

id., 'Medieval Herbals and their Development', *Clio Medica* 9 (1974), 23–33.

id., 'The theoretical bases of medieval herbalism', *Medical Heritage* 1 (1985), 186–98.

id., 'Folk Medicine, Philosophy and Medieval Herbalism', *Res publica litterarum* 3 (1980), 229–36.
id., *Selected Medieval Garden Plants and their Uses* (Lawrence, 1983)
Voigts, L., 'Anglo-Saxon Plant Remedies and the Anglo-Saxons', *Isis* 70 (1979), 250–68.

GARDENS

Cooper, G., *English Herb Gardens* (London, 1986)
Desmond, R., *Bibliography of British Gardens* (Winchester, 1984)
Fischer, H., *Mittelalterliche Pflanzenkunde* . . . pp.127–84
Genders, R., *The Cottage Garden and the old-fashioned Flowers* (London, 1969; 1983[2])
Harvey, J., *Mediaeval Gardens* (London, 1981)
Hyams, E., *The English Garden* (London, 1964)
id., *English Cottage Gardens* (London, 1970)
id., *A History of Gardens and Gardening* (London, 1971)
MacDougall, E. B. & Jashemski, W. (eds.), *Ancient Roman Gardens* (Dumbarton Oaks, 1981)
eadem & Miller, N., *Fons Sapientiae: Garden Fountains in Illustrated Books, Sixteenth-Eighteenth Centuries* (Dumbarton Oaks, 1977)
McLean, T., *Medieval English Gardens* (London, 1981)
Palliser, S. M., *The Use of Plants in English Folk Medicine 1600–1800* (Leeds, 1984)
Stokstad, M. & Stannard, J., *Gardens of the Middle Ages* (Lawrence, 1983)
Thacker, C., *The History of Gardens* (London, 1979; repr. 1985)

HERBALS (FACSIMILES)

Hunger, F. W. T., *The Herbal of Pseudo-Apuleius from the Ninth-Century Manuscript in the Abbey of Montecassino (Codex Casinensis 97)* . . . (Leyden, 1935)
Leporace, T. G., Pollacci, G., Maffei, S. L., *Un inedito erbario farmaceutico medioevale* (Firenze, 1952) [= Bibl. univ. di Pavia, Mss. Aldini 211]
Pfister, A., *De simplici medicina. Kräuterbuch-Handschrift aus dem letzten Viertel des 14. Jahrhunderts im Besitz der Basler Universitäts-Bibliothek* (Basel, 1960)
Gerstinger, H., Dioscurides Codex Vindobonensis med. gr. 1 der Österreichischen Nationalbibliothek 5 vols. (Graz, 1965–70)
Baumann, F. A., *Das Erbario Carrarese und die Bildtradition des 'Tractatus de herbis'* (Bern, 1974) [= MS London, B.L. Egerton 2020]
Lupo, M., *L'Erbario di Trento: il manoscritto n. 1591 del Museo Provinciale d'Arte* 2nd. ed. (Trento, 1978)
Herbarium Apulei 1481, Herbolario 1522 con una introduzione di Erminio Caprotti e un saggio di William T. Stearn, 2 vols. (Milano, 1979)
Melius, P., *Herbárium. Az fáknak, füveknek nevekröl, természetekröl és hasznairól* (Bukarest, 1979)

Pezzella, S., *I Segreti della medicina verde nell'epoca medicea da due manoscritti inediti della Città di Firenze (secoli xv e xvi)* (Assisi, 1980) [= MS Redi 165 of the Bibl. Mediceo-Laurenziana & MS 168/C of the Bibl. Marucelliana]
Werneck, H. L. & Speta, F., *Das Kräuterbuch des Johannes Hartlieb: eine deutsche Bilderhandschrift aus der Mitte des 15. Jahrhunderts* (Graz, 1980)
Minutelli, S.R., Lupo, M., Giulini P., *Erbario anonimo del xv secolo, Codice Marciano it. Z. 78 (= 4758)* (Venezia, 1980)
Ragazzini, S., *Un erbario del xv secolo. Il ms. 106 della Biblioteca di botanica dell'Università di Firenze* (Firenze, 1983)
P.O. Matthioli, Herbář jinak bylinář velmi užitečný (Praha, 1982)
Opsomer, C., *Livre des simples medicines: codex Bruxellensis IV 1024*, 2 vols (Antwerpen, 1980)
A. Laguna, Pedacio Dioscórides Anazarbeo (1555) 2 vols. (Madrid, 1968–69)

SOME INDIVIDUAL PLANT STUDIES

Brøndegaard, V. J., 'Wegerich als Wundheilmittel in der Volkes-und Schulmedizin', *Sudhoffs Archiv* 47 (1963), 127–51
Dieck, A., 'Malva silvestris, Capsella bursa-pastoralis [sic] und Juniperus communis in vor- und frühgeschichtlichen Funden', *Curare* 2 (1979), 117–24
Dodgson, J. McN., 'ME "cronebery" ', *Notes & Queries* 219 (1968), 88–89
Harvey, J. H., 'Mediaeval Plantsmanship in England: the culture of Rosemary', *Garden History* I, 1 (1972), 14–21
id., 'Gilliflower and Carnation', *ibid.* VI, 1 (1978), 46–57
Kärre, K., 'The English Plant-Name *groundsel*', *Studier i modern språkvetenskap* 9 (1924), 69–78
McLure, P., 'Three Plant Names in ME Place Names and Surnames: "Breme", "Rounce", "Bilberry" ', *Notes & Queries* 219 (1974), 42–44
Macht, D. I., 'Calendula or Marigold in Medical History and in Shakespeare', *Bulletin of the History of Medicine* 29 (1955), 491–502
Nordhagen, R., *Kusymre, kodriver, cowslip og paigle: studier over gamle Primula-navn i nordvest-Europa* (Oslo, 1955)
Stannard, J., 'The Multiple Uses of Dill (Anethum graveolens) in Medieval Medicine', *Festschrift Willem F. Daems*, Würzburger Medizinhistorische Forschungen 24 (Pattensen, Han., 1982), pp. 411–24
id., 'Squill in Ancient and Medieval Materia Medica with special reference to its employment for dropsy', *Bulletin of the New York Academy of Medicine*, sec. ser. 50 (1974), 684–713
Vermeer, H. J., 'Cardo benedicta das edlist Krautt' in G. Keil *et al.* (eds.), *Fachliteratur des Mittelalters: Festschrift für Gerhard Eis* (Stuttgart, 1968), pp. 421–32
Zimmermann, V., 'Der Rosmarin als Heilpflanze und Wunderdroge: ein Beitrag zu den mittelalterlichen Drogenmonographien', *Sudhoffs Archiv* 64 (1980), 351–70
van der Zweep, W., 'Linguistic, Artistic and Folklore Aspects of Tares in the

Biblical Parable' in R. Vickery (ed.), *Plant-Lore Studies* (London, 1984), pp.162–79
El-Gammal, S. Y., 'Garlic in History', *Hamdard Medicus* 28 (1985), 85–89

The following four articles appeared in the short-lived and rather inaccessible journal *Kyklos* (Jahrbuch des Instituts für Geschichte der Medizin an der Universität Leipzig):
Heyser, K., 'Die Alliumarten als Arzneimittel im Gebrauch der abendländischen Medizin', vol. 1 (1928), 64–102
Hirschfeld, E., 'Studien zur Geschichte der Heilpflanzen 1. Lilium convallium', vol.2 (1929), 145–62
id., 'Studien zur Geschichte der Heilpflanzen 2. Scilla', *ibid.*, 163–79
Tscholakowa, M., 'Studien zur Geschichte der Heilpflanzen 3. Verwendung des Safran (Crocus sativus)', *ibid.*, 179–90

PLANT NAMES OF MEDIEVAL ENGLAND

Aaron see Barba Aaron

Abaligustra
?Primula veris L. ? Cowslip
kouslope(B5)

Abies
Abies Mill. & other pinaceae Fir
fir-tre(C9)

Abitumon [albutume]
Allium ursinum L./*vineale* L. Ramsons, Wild Garlic/Crow Garlic
anglice wylde garleke(C6) anglice wilde garlec(C8)

Ablacta
?Daphne gnidium L. ?'Spurge Flax'
craunebene(C37)
Lit. B III, 61

Abrotanum [abrotonus, abrotinum, abrotonum, abratanum, abrotumum]
Artemisia abrotanum L. Southernwood
gallice averroyne, anglice suthernewod(A1) gallice averoyn, anglice suthernewode(A2) averoyne, suþernewodde(A3) gallice averoyne(A4) gallice averoyne, anglice souþernwode(A6) southrenewode(A7) averoyne, sowþernewode(A8) gallice averoyne(A9) anglice sozerneuuode(A10) gallice averoyne, anglice souþernewode(A12) gallice averoyn, anglice southerenwod(B1) gallice averoigne, anglice southernewode(B2) southernewode(B3) averoyne, southernwode(B4) gallice averoyne.i.suthernewode(B5) averoyn(B6) averoyne, soþronwode(B8)
anglice sowþernewode, gallice averoyne(C1) (a. agreste) wild sowþer(n)wode(C1) anglice averounce (C3) averougne(C7) (agreste) wild avoroyn(C26) gallice averoyne, anglice southerwod(C36) (agreste) þe lasse soþerwode(C38)

Absinthium [a. amarum, a. maius]
Artemisia absinthium L. Wormwood
gallice aloyne, anglice wermot(A1) aloyne, wermot(A3) anglice warmot(A4) gallice aloygne, anglice wermod(A6) wormod(A7) uermot(A10)
gallice aloyne, anglice wormode(B1) gallice aloyne, anglice weremod(B2) wormode(B4) wermot(B5) wormod(B6) aloyn, wormod(B8)
anglice wermod, gallice aloyne(C1) anglice wormote(C3) anglice wermode or symon(C10) wormewode(C28) wormewode, aloyne, symount(C31)

Absinthium Ponticum
Artemisia maritima L./*pontica* L. Sea Wormwood/Roman Wormwood
gallice centonicle, anglice sewormot(A1) sewermot(A3)
gallice centonicle, anglice weremod(B2)
gallice centonicle(C1) gallice centonike, anglice sewermod(C5) gallice sentenik, anglice seewormot(C6) gallice sentonyk, anglice sewermod(C8) gallice centonicle, anglice woremode(C18) gallice centonik, anglice sewermoth(C36)
Lit. Alph., p. 38, 11

Abusardan [abursardan, albusandran, ambusardan]
?*Dipsacus fullonum* L. ? Teasel
?*Dipsacus sylvestris* Huds. ? Wild Teasel
anglice wilde tasele(C13) wyld tasyll(C21)

Aca
?*Allium ursinum* L./*vineale* L. ? Ramsons, Wild Garlic/ Crow Garlic
anglice wylde garlek(B7)
anglice wylde garlec(C6) wyld gerlek(C36)

Acalcas
?*Geranium maculatum* L. ? Wild Geranium, Dove's-Foot
?*Geranium molle* L. ? Dove's-Foot Cranesbill
alion, culverfote(C21)
Lit. C8 = 'alion: aluina, sparagus, sparge'

Acaleas [acabeas]
'Sparagus'; variously applied (glosses error for sparge?)

anglice sporge(B7)
anglice spurge (C5, C6, C13)

Acalephe [achalice, acalisa, acaliphe, acalifa]
Lamium purpureum L. Purple Dead-Nettle
Urtica L. ssp. Nettle
netelles(B4)
anglice red-nettyll(C6) reed-nettil(C21)
Lit. André I, p.14 & II, p. 2

Acandus
?*Campanula* L. ssp. ? Bellflower
belwort(B2)
anglice belwort(C1) anglice belliwort(C16) belwort(C18)

Acanta Egyptiaca
?*Rhamnus catharticus* L. ? Buckthorn
?*Rubus* L. ssp. ? Bramble, Blackberry etc.
þewþorn(C25)
Lit. André I, p.14 & II, p. 2; Alph., p.1,12

Acanteleuca [aconteleuta]
?*Cnicus ferox* L. ? 'Cruel Thistle'
smale thistell(C28)
Lit. B III, 210

Acanton [acantus]
Melissa officinalis L. Common Balm
Onopordon acanthium L. Scottish Thistle
?*Acanthus mollis* L./*spinosus* L. ? Bear's Breech
?*Meum athamanticum* Jacq. ? Meu, Spignel
bewort(B3) (acantus) heiwort [corr. bewort?](B6) anglice berewort(B7)
anglice berwort(C5) anglice beoworte(C6) bewort, anglice wilde
garlek*(C8) kerwrt*(C13) b(l)ewwort(C21) bewrt(C23) beworth(C36)
Lit. B III, 23; N146

Acantum [ascantum]
Urtica dioica L. Nettle(seed)
semence de urtie, netylsed(A3)
gallice la semence de urtie, anglice netlesed(B2)
anglice nettelsede(C5) nettylsede(C21) nettelsed(C25) anglice grete netel
or blynd netel or bonewort(C26)

Accitorium
?*Allium ursinum* L. — ? Ramsons, Wild Garlic
anglice rammescrop(B1)
Lit. B III, 142

Accitula [atticula]
Allium ursinum L. — Ramsons, Wild Garlic
rammes(B4)
ramese(C28)
Lit. André I, p.15

Acelbar
?*Euphorbia lathyrus* L. — ? Caper Spurge
?*Daphne laureola* L. — ? Spurge Laurel
guþecorne(C25)
Lit. B III, 112f.

Acera
?*Centaurea scabiosa* L. — ? Great Knapweed, Ironhead, Hard Irons
?*Centaurea nigra* L. — ? Lesser Knapweed, Hardheads
Glechoma hederacea L. — Ground Ivy
irenhard(B3)
anglice heyhove or erth-yvy or grond-yvy(C26)

Acetosa
Rumex acetosa L./*acetosella* L. — Common Sorrel/Sheep's Sorrel
Oxalis acetosella L. — Wood Sorrel, Wood Sour, Stubwort
wodesoure(A3)
sourdocke, sorell(B4)
(a. minor) alleluia, anglice wodesowre, stobwort, cokkowsowre, gallice pains cucuze(C1) gallice payn de coccowe vel surel dé boys, anglice wodsour(C5) gallice payn de cockeyl vel sorel dé boys, anglice wodsowre(C6)

Acetula
Allium ursinum L. — Ramsons, Wild Garlic
anglice rammes(B1)
Lit. B III, 142

Achia [acha]
Apium graveolens L. — Wild Celery, Smallage, Water Parsley

smalache, water-persil(C9)

Achillea
Nymphaea alba L. — White Water-Lily
anglice colencrop(B1)
collcrop(C28)
Lit. B III, 57 (colloncroh)

Acidula [acedula, acridula, assedella, acerdula, acedola, acedilla]
Rumex acetosa L./*acetosella* L. — Common Sorrel/Sheep's Sorrel
Oxalis acetosella L. — Wood Sorrel, Wood Sour, Alleluia
gallice eysel, anglice surdoke(A1) esille, surele, surdokke(A3) gallice surele, anglice surdok(A4) sorel vel surel, anglice surdok(A5) gallice surele, anglice sourdoke(A6) gallice surele, anglice surdoke(A8) gallice sorel, anglice souredoke(A9) gallice sorel, anglice surdekke(A12) anglice surdocke(B1) gallice surel, anglice sourdok(B1) gallice osyle vel surele, anglice sourdocke(B2) sourdocke(B3)
(a. maior) anglice sowrdokke, gallice o[s]yle vel sorele(C1) anglice sowredoke(C3) gallice ocile vel surel, anglice sourdoke(C5) gallice ociel vel sorel, anglice sowredokke(C6) asille, surele, sourdocke(C22)

Acidula Maior
Smyrnium olusatrum L. — Alexanders, Horse-Parsley
gallice alisandre(B1)
anglice stanmerch(C20)

Acidula Minor
Sempervivum tectorum L. — Houseleek
gallice jubarbe, anglice sengren vel houslek(B1)
(minor) jubarbe, houseleke, syngrene, erwort(C13) seingrene(C28)
Lit. cf. Alph., p. 20, 12 & p.175

Acinum
Rubus idaeus L. — Raspberry
anglice hindeberie(B1)
hyndbery(C10) hindeberien(C23)
Lit. B III, 134

Acitellum
Allium ursinum L. — Ramsons, Wild Garlic
anglice ravenescrop [for rammescrop](B1)
ravenescrep[sic](C28)
Lit. B III, 142 cf. alternation rammesfot/ravenesfot

Aconitum [aconita]
?*Aconitum anglicum* Stapf — ? Monkshood
?*Doronicum pardalianches* L. — ? 'Great Leopard's Bane'
?*Aconitum vulparia* Rchb. — ? Wolf's-Bane
?*Digitalis purpurea* L. — ? Foxglove
?*Lolium temulentum* L. — ? Darnel
celerata, gallice devee(B1)
foxglove(C21) darnell(C28)

Acorad [alcorad]
Allium porrum L. — Leek
anglice leke(B7)
anglice leke(C5, C6) anglice lek(C8) anglice leek(C13) porret(C21) anglice lek or porrett(C26)
Lit. MS x (acorad)

Acorus [achorus]
Acorus calamus L. — Sweet Flag
Iris pseudacorus L. — Yellow Flag
gallice glejole, anglice segge(A1) racine de glogel, sech(A3)
la racyne de gladiole, anglice segge(B2) gladene(B3)
anglice segge(C1) gallice le racyn de glayole(C5) la racyn de gladiole(C8) gladne, slycewort or swerdgras(C9) þe rote of þe gladiole(C10) segge(C15) anglice gladen(C19) racine de glejole, sej(C22) anglice swerdwort or gladyn(C26)
Lit. N146f.

Acorus 2
?*Rubia tinctorum* L. — ? Madder
warence .i. madder(C7) mader .i. radix gladioli(C28)

Acride
Pastinaca sativa L. — Parsnip
passenepe(C10) pastnepe(C24) pastyrnepe(C27)

Acro Agrestis [acros agreste]
?
woode-whystill or kex(C33) anglice wodewhistel or kex(C37)

Acrocerium
Rumex L. ssp. Dock
Arctium lappa L. Burdock
gallice gletoner, anglice clot(B1)
clote(C28)
Lit. B III, 71

Actis
Sambucus nigra L. Elder
Sambucus ebulus L. Danewort
gallice saunbuk, anglice elre(B1) anglice elder-tre(B7)
anglice hellerne(C5) anglice elleryn(C6) anglice elderne(C8) pety elder, walwort(C9) ellern tree(C10) eldern tre(C11) ellen, eldren (C13) ylder(C24) helrene-tre(C27) eldren(C28)
Lit. N147

Acumusca [for acus muscata?]
?*Hyoscyamus niger* L. ? Henbane
henbane*(C21)

Acus Demonis [a. demonum]
Erodium moschatum (L.) L'Hérit. 'Musk Storksbill'
Scandix pecten-veneris L. Shepherd's Needle
puk-nedel, styke-pilor(C1) anglice powke-nedill, stike-pyle(C4) pouke-nedle, styke-pile(C32) pouk-nedyl(C38) pouke-nedyll or styke-pyle(C41)
Lit. cf. Alph., p. 2, 25 (acus muscata); Agnus Castus p. 212

Acus Diaboli
Filago germanica (L.) L. Cudweed
anglice horwort(B7)
anglice horwort(C5, C8) horworth, chaunette(C13) herb Robert*, horwort(C21)

Acus Muscata [a. mustula]
Erodium moschatum (L.) L'Hérit. 'Musk Storksbill', 'Crowbill', 'Herb Robert', 'Ground Needle Storksbill'
Geranium L. ssp. esp. *G. robertianum* L. Herb Robert

cropil(A3) anglice crouenedles(A8)
anglice croubile(B1) (maior) herbe Robert (minor) mout est sanatif, anglice crouwepil(B2)
(minor) erþenedul, crowudnedul(C1) anglice crowpil vel crownedel(C5) (minor) anglice crowlef vel crousope*(C8) Salomonys trone (tronus Salomonis), þe grete tronarye (tronaria magna), þe grete spurge, crownedlen et chirsnedle(C9) (maior) herbe Robert (minor) croppille(C10) croweeppil, crowenedell(C13) (maior) herbe Robert(C16) (a. minor) mout sanatif, anglice crouwebil(C18) crownedyl(C21) (maior) herbe Robert þe more(C26), (maior) herb Robert(C24) herb Roberd(C27) (minor) crowpile(C27) ramys*(C29) croweneldus(C32) develes nedle(C35) erþenedyl(C38) (minor) anglice grounde-nettyll [corr. nedyll](C41)

Adactis
?*Lonicera periclymenum* L. ? Honeysuckle
?*Tragopogon pratensis* L. ? Goat's-Beard
?*Convolvulus arvensis* L. ? Bindweed
wodeberd (barba silvana), gotysberd (barba hyrci), wodebynde (liga silve)(C9)

Adarasca [adarascus, adarasci]
Veratrum album L. White Hellebore
Helleborus viridis L. Green Hellebore
seterwort(B3)
(eleborus aquatica) water elebre, sikelwort(C9) anglice seterwort(C12) elebre(C21) anglice longewort(C26)
Lit. N147

Adaserta
?
wodwrte (MS. Cambridge U.L. Dd.5.53 f.2ra)

Adhibenda [adibunda]
Capsella bursa-pastoris (L.) Medic. Shepherd's Purse, Sanguinary
anglice stawncheblode(B7)
anglice stancheblode(C8)

Adiacus
?
foxglove, cantabel(C35)

Adiantos [adiantes, adiantis]
Adiantum capillus-veneris L. Maidenhair-Fern
Asplenium adiantum-nigrum L. 'Black Spleenwort'
Asplenium trichomanes L. 'Maidenhair Spleenwort'
madenhere(B4)
maydenher(C9) sparhawkysfote (pes nisi)(C10) papwort(C13) saxifrage(C21) sperhaukesfote(C24) maidenhere(C24) þe more maydenhere or netilwort(C26) sperhaukefote(C27)
Lit. André I, p.18 & II, p. 5

Adigitone
Arum maculatum L. Cuckoo-Pint
panis cuculi, anglice skabewort [corr. stubwort?](C37)

Adriaca [adriatica]
?*Symphytum officinale* L. ? Comfrey
?*Allium sativum* L. ? Garlic
kerwrt(C13) garlek(C28)
Lit. B III, 101–2

Affodilla
Asperula odorata L. Sweet Woodruff
muge, anglice wodrowe(C16) anglice woderove(C43)
Lit. B II, 132

Affodillus [afodulus, afidoli, affrodilis, affodilla, affrodillus, fodissus]
Allium ursinum L./*vineale* L. Ramsons, Wild Garlic/Crow Garlic
Asperula odorata L. Sweet Woodruff
anglice wodhove vel wylde garlek(B1) herba est manere de noys [sic], anglice wilde garlek(B2) anglice clote . . . anglice ramesen(B2) anglice ramese(B5) croulek(B6)
rameson(C1) affodille, sent-chef, clote(C1) anglice affadill(C4) wodehove(C7) wilde garlyk(C13) rammes(C15) anglice clote(C16) (asfodillus) anglice clote (affodillus) rameson(C18) ramsys(C21) centum capita, syn-chef(C27) wilde leke(C28) affadille(C35) wodehove(C37) anglice affodyll, gallice centibef [corr. cent-chef](C41)

Affricanus [crocus a.]
Crocus sativus L. Saffron
anglice safron(B7)
anglice safron(C6) anglice saffran(C8) safferan(C13) saffron(C21) safron de orte(C21) safron de ort(C27)

Affrodite
Dipsacus fullonum L. Teasel
Dipsacus sylvestris Huds. Wild Teasel
(?) .i. locher .i. cardus maculatus(C1)
Lit. André I, p. 34 (Aphrodites lutron)

Affros
Sanicula europaea L. Wood Sanicle
senecle, wodmarche(B4)
wodemerche(C4) sanicle, wodemerch(C7)

Afida [for favida?]
?*Veronica beccabunga* L. ? Brooklime
lemeke(C28)

Afrisa
Fragaria vesca L. Wild Strawberry
freses(A11)

Agallus [agallis, agalla]
Agrostemma githago L. Corn Cockle
Lolium temulentum L. Darnel
anglice cokkul(B7)
anglice cockel(C5, C6)

Agaricus
Polyporus officinalis L. Agaric, Larch Agaric
agarif [corr. agarik?](A3)
fungus abietis, musseron(B3)
mosseron(C2) agaryf [corr. agaryk?](C22) agarik(C31)
Lit. N148f.

Agiros [agiro]
Populus L. ssp. Poplar
pople-tre(C21) anglice popler-tre(C26)

Agnus Castus
Hypericum androsaemum L. Tutsan
?*Hypericum perforatum* L. ? St. John's Wort
Vitex agnus-castus L. Chaste Tree
'*Salix marina*' ?
quintefoille(A7)
tutseyn(B3)

andra maior et minor, totsayne þe more and þe lasse, al-hele, parcleves(C1) gallice toutsain(C2) tutsayn, pollitryk(C13) anglice selfhele(C16) toteseyn(C24)salix marina, anglise seewyþy vel bisschoppeswort(C5) anglice sorwythy vel bysschoppusworth(C6) anglice sewethi vel bisshopeswort(C8) bisshopeswort(C9) agnes cast or bysshoppeswort(C21) quintefoyle(C22) tutsayn or bysshopswort(C26) tutsane(C28) chastane*(C29) tutseyne, byschopyswort(C31) se-welȝe(C35) anglice totsane or pentifelon(C41)

Agon
?*Adiantum capillus-veneris* L. ? Maidenhair-Fern
?*Asplenium adiantum-nigrum* L. ? 'Black Spleenwort'
?*Asplenium trichomanes* L. ? 'Maidenhair Spleenwort'
sparowisfote(C24)

Agradiadema [argadiademe, argiadiadema, agrediadema, angreduldema, agradiana]
Glechoma hederacea L. Ground Ivy
hofe(C10, C11) hove(C7, C13) hofe(C27) a hest[?] hove(C40)

Agramen
?
(cordopacea, tubara) anglice everwort(C26)

Agrimonia [acrimonia, argimonia]
Agrimonia eupatoria L. Agrimony
egremoyne(A1, A3, A8) egrimoyne(A6)
anglice clefwort, gallice egermoyne(B1) gallice egremoigne(B2) egrimoyne, garscleve(B3) egrimoyne, garclyve(B4) egrimoine(B5) egremoyn(B6)
egermoyne, agrimoni, coseclite(C1) artennale, argemo, artelan, coseclite vel agrimonia(C1) anglice agrimonde(C3) egrymoyne, garsclyve(C10) anglice egrymoyn or appelwode*(C26) egrimony, gars cheline [corr. garsclive](C27) clifwede(C28) garclive(C37)
Lit. N149

Agrimulatum [agrimolatum, acrimulatum]
Agrostemma githago L. Corn Cockle
Lolium temulentum L. Darnel
kockel(A3)
cockel(B2) cokkel(B3) anglice cockul(B7)
cokkel(C2) anglice cockyl or darnell(C26)
Lit. N147

Agrion [for agrion cannabum?]
Eupatorium cannabinum L. — Hemp Agrimony
Cannabis sativa L. — Cannabis
anglice hempe(C26)

Agripmellus
?*Anacyclus pyrethrum* DC. — ? Pellitory of Spain
?*Allium porrum* L. — ? Leek
petir(C35)

Aiacon
Adiantum capillus-veneris L. — Maidenhair-Fern
Asplenium adiantum-nigrum L. — 'Black Spleenwort'
Asplenium trichomanes L. — 'Maidenhair Spleenwort'
maydenher(C25)
Lit. Alph., App. p. 201 ('aracon .i. capillus veneris') & N149

Aiarisalon
?*Aristolochia rotunda* L. — ? Aristolochia, Birthwort
aristologie(C25)
Lit. cf. André I, p.18 (adra riza) & p. 269 (raia); N154 (araziza)

Aison [ayzon, azios maior & minor]
Rumex acetosa L./*acetosella* L. — Common Sorrel/Sheep's Sorrel
Oxalis acetosella L. — Wood Sorrel, Wood Sour, Alleluia
anglice sowredok(B7)
anglice sourdok(C5) anglice souredocke(C6) docke(C28) (maior) wodesure (minor) sourdok(C28)

Aizoon [aysone, azon, aycon]
Sempervivum tectorum L. — Houseleek, Sengreen
?*Adiantum capillus-veneris* L. — ? Maidenhair-Fern [conf. w. alacon?]
jubarbe, erworte(B4)
jubarbe, herewort, synegrene(C10) maydenher*(C2, C12) jubarbe, herworth, senygrene, housleke(C24) jubarbe, herwort, selvegrene(C27) jubarbe, herworte, selvegrene(C40)

Alabum [alibuum]
Urginea maritima (L.) Baker — Squill, Sea Onion
squillis(C10) squielles(C11) squilles(C27)

Alacon [alacen, aleten, alcen, alocon]
Adiantum capillus-veneris L. Maidenhair-Fern
Asplenium adiantum-nigrum L. 'Black Spleenwort'
Asplenium trichomanes L. 'Maidenhair Spleenwort'
anglice maydenher(C5, C8, C11) anglice maydenhere(C6)
Lit. N149

Alapsa
Quercus L. ssp. Oak Gall, Oak Apple
gallys, ok-appul(C9) anglice oke-apples or oke-galles(C26)

Albatan
?*Caesalpinia* L. ssp. ? Brazilwood
brasyle(C25)

Albida [albita]
?*Silene* L. incl. *dioica* (L.) Clairv. 'Flos campi', ? Red Campion
(maior) flos campy, synagre(C9)

Albifa [albipha]
?*Mandragora officinarum* L. ? Mandragora, 'Satan's Apple'
þe appel of þe mandrag(C10) þe appyl of mandrake(C11)

Albor
?*Coriandrum sativum* L. ? Coriander
coriandre(C10) coriander(C27)

Albucium [albitumen, albutron]
Allium vineale L./*ursinum* L. Crow Garlic/Ramsons
?*Asphodelus ramosus* L. ? Asphodel
allea corvorum, crowesgarlek(C21) affodillus, affodille(C21) wild gerlec(C36) anglice ramesen(C41)
Lit. André I, p. 22 & II, p. 9

Alcanna
Alkanna tinctoria (L.) Tsch. Alkanet
?*Lawsonia inermis* L. ? Henna
?*Pentaglottis sempervirens* (L.) Tsch. ? Alkanet
(a. maior) alkenete, blite (a. rubea) red alkenete(C1) alkenade(C31) alkette [sic](C32) alkenet(C37)

Alcia
?*Lavandula spica* L. ? Spike Lavender

?Valeriana celtica L. ? Celtic Spikenard
þe grete shergres spik(C9)

Aleos
'Aloes epaticum', 'aloes hepatites' form of Socotrine Aloes
epatik(C11)

Aleptafilus (lepthapilos)
Artemisia vulgaris L. Mugwort
Artemisia campestris L. Field Southernwood
anglice mugwort(C19) gallice ermoyse(C20)
Lit. André I, p.184 (leptophyllos) & II, p.142

Alexandrea [alexandrum]
Smyrnium olusatrum L. Alexanders, Horse-Parsley
stanmarche, genciane, alysandir(B4)
alisaundir, stanmerche(C4)

Alexandrina [for cucurbita alexandrina]
Citrullus colocynthis Schrader Colocynth
?Lagenaria vulgaris Ser. ? Gourd
gourd(C25) stammiryngeswort(C37)
Lit. André I, p.107 & II, p. 80; Alph., p. 46, 19

Alfel
Sambucus nigra L. Elder
anglice elder-tre(C26)

Alga [algea]
Algae esp. *Glyceria fluitans* (L.) R.Br. Alga esp. Flote-grass
anglice sewor(A2)
gallice her de fey, anglice leverwort*(B1) feltworte [corr. fletworte](B4) sewort(B5)
flotwort(C7) flotwort(C10) sewor(C10) fletwort(C11) sewrt(C13) flotewort(C16) sewor(C27) lyveruett*(?)(C28) flootwort(C40)
Lit. B III, 96

Algachir
?Mercurialis perennis L./*annua* L. ? Dog's Mercury/'Annual Mercury'
mercury(C25)

Algon [algol]
Adiantum capillus-veneris L. Maidenhair-Fern
Asplenium adiantum-nigrum L. 'Black Spleenwort'
Asplenium trichomanes L. 'Maidenhair Spleenwort'
sperhaukesfote(C10) sparhaukefot(C11)
Lit. N149

Alguason
Iris Florentina L. Florentine Iris
anglice flour de lice(C26)

Alimonus
Papaver L. ssp. Poppy
wyld popy(C25)
Lit. cf. André I, p.156 (halimon) & II, p.116

Alipiades [alipiados, alypias]
Daphne laureola L. Spurge Laurel
laureall(B4)
laureale(C3) laureole(C10) laurel(C21)
Lit. N150

Aliptis [alipti]
Brassica rapa L. Turnip
anglice nepe(B7)
anglice nepe(C5, C21) anglice nepte(C6) anglice nep(C8, C13) anglice hape [corr. rape?] or nepe(C26)
Lit. A. App. 204 (alipti)

Aliricum
?*Ligusticum scoticum* L. ? Lovage
?*Levisticum officinale* Koch ? Lovage
loveache(C21)

Alisander
Smyrnium olusatrum L. Alexanders, Horse-Parsley
stanmerche .i. wilde persile(C35)

Alithimus
Cinnamomum zeylanicum Blume Cinnamon
canel-tre, (alithimum) þe frut, þe rynde of canel-tre(C9)

Alitum
? ?
klete(B5)

Alkekengi [alkakenchy, alkicakengi]
Physalis alkekengi L. Winter Cherry
conf. w. *Solanum nigrum* L. 'Black Nightshade'
þe myddel morell(C1) anglice myddill morell, anglice nyghtshode, mixbere(C4) anglice alkyn(kyn)gyngy(C26)

Alkenet
Atriplex L. ssp. Orache
on of þe aragys . . . þe erbe þat þe comown calleþ arage . . . proprely for þe grete arage (alkenet enytron)(C9)

Alkubarub [altubarub]
Satureia hortensis L. Savory
saverey(C21)
Lit. C8 = altubarub herba similis ysopo et secundum quosdam dicitur satureya

Alleluia
Oxalis acetosella L. Wood Sorrel, Wood Sour, 'Cuckow's Meat', 'Gowke Meat'
gallice payne cucu, anglice yekesmete .i. wodesour(A1) pain de cucuil, coccospit, yekesmete .i. wodesour(A3) wodesoure(A3) anglice wodesoure(A6) wodesoure(A7) anglice weudesure(A8) anglice wodesure(A9) wodesour(A12)
wodesoure(B1) gallice payn cucuze, anglice wodesoure(B2) wodesoure(B3) sorell dé boys, wodesoure, stobwort, gowkeshore(B4) anglice wdesure(B5) gallice surel, anglice goukesour(B9)
anglice alleluia vel wodesowre(C2) anglice wodesowre or gokesowre(C3) wodesour, stubwort(C4) gallice payn de cocku vel sorele dé boys, anglice wodesour(C8) sorel dé boys, anglice stubworte(C12) gallice pains cucuze, anglice wodesour(C18) cockespyt, wodesoure(C22) wodesour, chemerek, stevenwort(C24) anglice sorel de boyes or wodesour or cuckowort(C26) payn de cuculle, wodsour, stubwort(C27) stubwort, woode-swythe(C31) coccusoure(C35) anglice coksoure, sobbewort, wodsoure, stubwort(C41)

Allium Agreste [allia a.]
Allium ursinum L. Ramsons, Wild Garlic

Allium vineale L. Crow Garlic
anglice ramese(A6)
gallice auz, anglice garlek(B1) crowgarlek(B3) crauleche(B5)
(a.a., silvestre, minus) wild-, crow-, lasse-garlyk(C1) crowegarlek(C2)
anglice crakeleke(C3) anglice crowgarlok [sic] or wilde garlek(C4)
(silvestre) rameson, wylde garlek(C6) (silvestre) ramesyn(C13) anglice
crakelek(C20) wilde leke(C35) crowesgarleke(C39)

Allium Domesticum
Allium sativum L. Garlic
gallice ayle, anglice gorlec(A1) ayl, garlec(A3)
deu manere de ausce, gallice ayle, anglice garlek(B2) garlyke(B4)
(commune) garlek(B3)
(a.d., maius) þe more garleke, gallice ayle(C1) gallice ail vel auce,
anglice garlek(C6) tame garlek(C9) seynt Mari gerleke(C35)

Allogalia [alogollia]
Gentiana L. ssp. esp. *lutea* L. Gentian
genciane(B3)
genciana(C25)
Lit. André I, p. 25

Allogallica [allusgallica, allocallica, alla gallica, allium gallicum]
Gentiana L. ssp. esp. *lutea* L. Gentian
gencian(C10) gentyan(C21) baldemoyn(C24) baldemune(C35)
baldemoyne(C36)
Lit. André I, p. 25

Alochias
a prickly plant, caltrops
kalktrap(C28)

Aloes
Aloë L. ssp. esp. *succotrina* Lam. Aloes esp. socotrine aloes
aloen(A3)
aloes epatyk, for medicines; aloes caballinum for plaistres onely(C21)
aloyn(C22) liverwort(C35)
Lit. André I, p. 24 & II, p.11

Alon
Salix L. ssp. Willow
salw(C13)

Aloser
?*Thymus serpyllum* L. ? Thyme
tyme(C25)
Lit. cf. MS xxiiii ('alasor .i. ysopus ortulanus')

Alotis
Gentiana L. ssp. Gentian
allium gallicum, genciane(C1) anglice gentiane or baldmony or bladewyne or fyldeworte(C26)

Alpha(Avis)
?*'semen fraxini'* ? Ash-key
?*Stellaria holostea* L. ? 'Greater Stitchwort'
gallice lang de oysel, anglice folestong(B1)
stughwert(C28)

Altercus
Hyoscyamus niger L. Henbane
henbayne(C3)
Lit. André I, p. 25 & II, p.11

Alteus
Artemisia absinthium L. Wormwood
gallice aloyn(B7)
succus cuiusdam herbe, gallice aloyne(C5) gallice aloen(C8)

Althaea [altea, altia]
Althaea officinalis L. Marsh Mallow
?*Althaea rosea* (L.) Cav. ? Hollyhock
mar[c]h [. . .], wymawe(A3) wymauve(A5)
gallice wymawe, anglice wysmalf(B1) treis maner de malve, gallice wimauve, anglice mersmalewe(B2) holyhocce(B3) holyokke, malowe, wymawe, march malwe(B4) gallice bismalve, anglice holihok(B7) (a. maior, rubea) þe more holyhokke (a. minor, alba) þe lasse holyhokke(C1) wymawe(C3) holihokke or wilde malow(C4) anglice holyhok vel merschmalue(C5) wymalve, holyhoc, holy malue(C7) whit malue(C13) anglice wyld malloue or seehocke(C26) holy malow(C29) se-hok(C31) French malue(C35)

Amabila
?*Atriplex hortensis* L. ? Orache
attriplex, comyn or melde(C21)
Lit. cf. B III, 157f.

Amantilla [amentilla]
Valeriana officinalis L. Valerian
valeriane(B3)
valerian(C4) valaryane(C21)

Amaracus [amaraca, amaricon, amariscus]
Majorana hortensis Moench Sweet Marjoram
Origanum vulgare L. Marjoram
majoran(C9) muchewort(C10) mugwort(C27)
Lit. André I, p. 26, N151

Amarantum
Artemisia vulgaris L. Mugwort
anglice mogwort(C1)

Amaras
Centaurium erythraea Rafn Centaury
Blackstonia perfoliata (L.) Huds. Yellow-Wort
centorie þe lasse(C1)

Amaricus
?
lathwort(C34)

Amarusca [amarisca, amariscus, amerusca]
Anthemis cotula L. Stinking Camomile, Mayweed
anglice maythe(A2) amertik [corr. amerok], maythe, hundesfenyl(A3) anglice hundefenil vel meythis(A5) anglice mawere(A6) anglice maythe(A8) anglice mayte(A9) anglice mayȝe(A12)
gallice ameroche, anglice maithe(B2) maithe(B3) anglice bitterwode vel maythe(B5) maythez(B6) gallice amaroc, anglice maythes(B9)
(a. maior, alba) mayþe, doggisfenl(C1) maythes(C2) maythe, filis et camamille(C3) dogfenell(C4) gallice ameroc, anglice mayth vel houndesfenkel(C5) maythe, dogfynell(C7) amaroke, maythe, houndesfenel(C13) gallice amaroche, anglice maythe vel bitterwort(C16) amentyk [corr. ameruk?], carthe(?), houndfenell(C22) maydewed(C23) amerok, hundefenell, matyn(C24) amerok, houndz-senemayne(C27) mawþene(C31) ameroc .i. hundesfencle(C35) maþe vel dogfenel(C38) maydwed(C39) anglice mawth, dogfenelle, maydynwod(C41)

Ambila
Allium vineale L. Crow Garlic

clove-leke [corr. croue-leke?](C28)
Lit. B III, 157f.

Ambrosia [ambrosium, ambros]
Teucrium scorodonia L. Wood Sage
gallice ambroise, anglice hyndehale(A1) ambroyse, hyndehale(A3) anglice lyverwourt*(A6)
gallice ambroyse, anglice hyndehele(B1) gallice ambrosie, anglice hyndehale .i. wilde sauge(B2) ambrosie(B2) wilde sauge, bryndhale, hyndhale(B4) ambrose(B5)
anglice wylde sauge(C3) gallice ambrose, anglice wilde sauge(C5) wo[d]sauge(C20) anglice hyndehale(C23) bindbal[sic] vel hendhale, wild sauge, ambrose(C24) .i. hynhole .i. wild sage(C35) byndehale[sic], wylde sauge(C40)

Ambrosia Minor
Achillea millefolium L. Yarrow, Milfoil, Nosebleed
anglice millefoyl, noseblede(C4) anglice ȝarowe(C5) noseblede(C6) anglice ȝarowe(C6) ȝarwe, noseblede(C31)

Amdo
?*Symphytum officinale* L. ? Comfrey
confirie(C28)

Ameos Agreste [amyos a., ameus, amea, ameys]
var. umbelliferae:
Cicuta virosa L. Cowbane
Anthriscus sylvestris (L.) Hoffm. Cow Parsley, Keck
Angelica sylvestris L. Wild Angelica
Conium maculatum L. Hemlock
Aegopodium podagraria L. Goutweed, Bishop's Weed
anglice wodewistel vel kex(A2) anglice wodeuistille(A6) wodequistle vel kex(A8) anglice wodewistel vel kex(A9) wodewistel anglice vel kex(A12)
wodewhistle(B3) wild wistil vel kex(B9)
(a.a. minor) kyxys(C1) anglice wodethystyll*(C3) anglice wodethestell* vel box*(C5) anglice wodeþystel* vel vex*[sic](C6) in some countre it is called kex, in some contre wodewhistel(C9) anglice wodewystle(C12) wodewiste, kex(C15) wodethistell*(C21) wodethistell* or box*(C21) anglice wodethistill*(C28) skirwhite*(C33) wodþistel* vel kex vel wild carewey(C36) widethistel*(C40)
Lit. B III, 259

Ameos Maior [a. ortulana]
?*Aegopodium podagraria* L. ? Goutweed
?*Ammi majus* L. ? Bishopsweed
amyas þe more(C1) amys(C42)

Ameratis [amerestis, amaracis, amaristis, amoracis, amorastis, ameratis]
Cynoglossum officinale L. Hound's Tongue
anglice houndestong(B7)
anglice houndestong-leve(C5) anglice hondestong(C6) houndistong(C21)

Amigdala
Prunus amygdalus Batsch Almond
Prunus dulcis (Mill.) D.A. Webb var. dulcis/amara
alemand(A3)
gallice almandes(B1) gallice et anglice alamande(B2)
almaundes(C2) anglice swete almondis, bitter almondis(C26)

Amomum
Papaver L. ssp. Poppy
popy(C28)

Amonedos
Artemisia vulgaris L. Mugwort
moderwort, hermeise(C37)

Amorfolia
Galium aparine L. Cleavers
?*Arctium lappa* L. ? Burdock
gallice reblet, anglice clevere(B1)
Lit. B III, 50f & 52; André I, p. 26

Amorsia [for amorfolia or amorosia?]
?*Arctium lappa* L. ? Burdock
?*Galium aparine* L. ? Cleavers
clotte(C25)

Ampelos [ampeleon, ampelon]
Bryonia dioica Jacq. (White) Bryony
?*Clematis vitalba* L. ? Traveller's Joy
wilde nepe(C9) anglice a wylde nepe(C12) wilde nep(C13)
Lit. André I, p. 28 (ampelos agria)

Amphora
?*Nuphar* Sm. ssp. ? Water-Lily
edhok(C21) edocke(C23)

Amphoricon [ampharicon]
Hypericum perforatum L. St. John's Wort
seynt Jon wort(C10) seynt John worte(C24) anglice erbe John(C39)

Ampleon
?*Urtica* L. ssp. ? Nettle
nettle(C25)

Ampullata
?*Phyllitis scolopendrium* (L.) Newm. ? Hart's-Tongue Fern
hertestonge(C14)
Lit. cf. André I, p. 29 (ampullaçia)

Anabulla
Euphorbia lathyrus L./*esula* L. (Caper) Spurge
Anthemis cotula L. Stinking Camomile, Mayweed
spurge, clansing-gras(A3) anglice holihokke*(A6) spurge(A8)
gallice bleuewode, anglice clansing-gras(B2) spurge(B3) spurge(B5)
gallice cataputz, anglice spurge(B9)
spurge(C1) spowrge(C3) spurge, dog-fennel, maidewode(C4) gallice beuerwode, anglice clansyng-gras(C16) gallice bleuerwode, anglice clansyng-gras(C18) spurge, clansynge-gras(C22) bleuewode, anglice clansyn[g-gras](C23) anglice spurgewort or ty(n)timal(C26) wylde spurge(C29) spurge, qua(r)terspys(C31) anglice faturgres(C36) (minor) gallice et anglice veroine[sic] or wrotwort*(C41)

Anacardia [anacardi]
Semecarpus anacardium L. Marking Nut
Anacardium orientale L. Anacard
anacardes(A3)
gallice anacardes(B7)
gallice anachardes(C6) þe lous of an elephant, and summe seyþ þat hit is þe fruyt of a tree(C10)

Anacarus [anachoris]
?*Sanicula europaea* L. ? Sanicle
sanicle, wodemerch(C24) sanicle, wodemerche(C40)
Lit. see **Ancorus**

Anacocci [anachorti, anachota, anachotus]
Laurus nobilis L. (Bay-) Laurel
bays(C2) þe grete lorer(C9) anglice bays(C12) anglice bayes(C26)
Lit. Alph., p.10, 12

Anadonia [anodina, anadonea, avadoma, anodonia, anodenia]
Verbascum thapsus L. Mullein, Feltwort, Aaron's Rod
anglice feldworth(B1) bischopiswort(B2) anglice bishopswort(B7) bischopiswort(C1) anglice bisschopeswort(C5) felterwort(C7) moleyn, anglice veltwort(C16) bischoppiswort(C18) bisshoppiswort(C21) ffeldwort(C28) felwort(C37) walwort*(C39) moleyn, feltwort(C43)
Lit. B II, 44f & III, 90

Anagalidos [agalido]
Cytisus scoparius L. Seed of Broom
Seed of Nettle*
bromsede(B6)
nettylsede*(C21) þe sede of nettyll*(C27)
Lit. MS xli ('semen mirice'); A. App. p. 203 (agalisis . . . semen urticae)

Anagallicum [anagallus, anagalla, anaglacium, anagallis]
Symphytum officinale L. Comfrey, Consound, Bruisewort
la grant conseude, conferye(A3)
anglice swynesnep*(B1) treis maner de consoude, gallice la grand consoude, anglice confirie(B2) confere(B3) brosworte(B4) gallice grant consound, anglice cunfery(B7)
bruswort(C2) gallice grant consoude, anglice confiri(C5) þe grete anagal, þe grete consoud, þe confirye(C9) anglice comfery vel brusewort(C12) the gret dayseye or confery(C21) the rote of confery(C21) þe rote of hym is clepyd simphitum, comfirie(C27)

Anagallus [amagallus, anagallis]
Hieracium pilosella L. Mouse-Ear Hawkweed
anglice mousere(B7)
auricula muris, moushere(C1) anglice mousere(C5) anglice moser(C6) consolida maior, dayeshie*(C30)
Lit. cf. André I, p. 30, L73

Anaspargas
Anthriscus sylvestris (L.) Hoffm. Wild Chervil, Cow Parsley
?*Asparagus officinalis* L. ? Asparagus
asparagus agrestis, wodefylle(C1)
Lit. B II, 131f & III, 265f.

Anatrina
Lemna minor L. — Duckweed, Duck's-Meat
anglice dokemete(C26)

Anaxilis
?*Symphytum officinale* L. — ? Comfrey
?*Bellis perennis* L. — ? Daisy
anena(?), the lesse consowd(C21)
Lit. André I, p. 29 (anachylis) & II, p.15

Ancorus [anchora, anacorus, anochorus]
Sanicula europaea L. — Wood Sanicle
sanicle or wodemerche(C10) canicle, wodemerche(C11) sanicle, wodemerkh(C13) sanycle or wodemerche(C21)
Lit. Alph., p.11, 11

A[n]cusa
'saliunca' — a prickly plant, caltrops
kaltrape(C19)
Lit. André I, p. 30, L70

Andra
?*Betonica officinalis* L. — ? Betony
bisschoppiswort(C20)

Andrago [andago, andrego]
Portulaca oleracea L./*sativa* L. — Green/Golden Purslane
purslane(B4)
purceleyn(C7) andrage, portulat, folefot(C9) grace gelyne*(C13) porceleyne(C17) purslane(C21) parcelane[sic](C27)

Anemone [anemonia, anemon]
Papaver L. ssp. — Poppy
red popy(B3)
popy(C9)
Lit. MS xliiii ('rubeus papaver')

Anesum
?*Sinapis arvensis* L. — ? Charlock, Wild Mustard
carloc(C9)

Anetum [anetus]
Anethum graveolens L. — Dill

Pimpinella anisum L. Anise
gallice anet(A1) anete, dile(A3) gallice anet, anglice dille(A4) anise(A8) anis(A11)
gallice anys, anglice dous [. . .](B1) gallice anete, anglice tille(B2) aneise(B3) dile(B5) anete, dile(B6)
anete, dylle(C1) anete(C2)

Anetum Agreste
Peucedanum officinale L. Hog's Fennel
hundefenel, wilde anet(C9) anglice beschopiswort(C16) anglice mathegyn or mayde or dog-fenel(C26)

Anicula
Allium ursinum L. Ramsons, Wild Garlic
ramese(C7)

Anisum [anisium, anisitta]
Pimpinella anisum L. Anise, Aniseed
gallice anis(A1) anise(A3) gallice anise(A4) anglice anys(A6) anglice anise(A8)
gallice anyse(B2) anise(B3) annes(B4) anice(B5) anyz(B6)
anise(C1, C2)

Anisus
Laserpitium siler L. Sermountain
Pimpinella anisum L. Anise, Aniseed
(maior) surmontey[n], (minor) anys(C9)

Annotana [annotona, annocana, anatana, anochana]
Bryonia dioica Jacq. (White) Bryony
?*Clematis vitalba* L. ? Traveller's Joy
wild nepe(B4)
anglice wild nepe(C4) wilde nepe or wilde vyne(C10) wylde neth [corr. nep], wylde vyne(C11) wild nep, wild vyne(C24) wyld nepe, wyld vyne(C27)

Annuosa
Sedum acre L. Common Stonecrop, Wall-Pepper
Sedum telephium L. Orpine, Stonecrop
gallice petit orpyn vel test de soriz, anglice stoncrop(B1)
stonecrop(C28)

Anta [for acantan?]
?*Carlina vulgaris* L. ? Carline Thistle
efortrode(C28)
Lit. B III, 81

Antadulla
?*Rumex acetosa* L./*acetosella* L. ? Common Sorrel/Sheep's Sorrel
anglice souredok(C43)

Antedron
Althaea officinalis L. Marsh Mallow
holyhok(C1)

Antera [anteria, anteta]
Rosa L. ssp. anther or pollen-bearing organ of the rose
gallice antere, la jaune flour dedenz la rose(B2) gallice anter(B7) gallice anter(C5) semens de rose(C7) rede rose(C9) antre, la jaune flour de rose(C16) gallice antere, la jaune flour dedens rose(C18) the seed of rose(C21)
Lit. N154

Anterogon
Hypericum perforatum L. St. John's Wort
herb John(C21)
Lit. C8 = anterogon: herba sancte Johannis

Anthemis [antemus, antonus]
Chamaemelum nobile (L.) All. Common Chamomile
camemille(C2) camamill(C14) camamylle(C21, C25)

Antheos
Chamaemelum nobile (L.) All. Common Chamomile
camamylle(B4)

Anthyllium [antilios]
Allium ursinum L. Ramsons, Wild Garlic
anglice ramese(B1)
Lit. cf. André I, p. 32

Antifarmacum [antifomatum, antiformatum, anty, antiframatum, anthipharmarum]
Aquilegia vulgaris L. Columbine

Potentilla erecta (L.) Räusch. Common Tormentil
Geranium L. ssp. esp. Columbine
G. columbinum L.
anglice rosegresse(B7)
anglice rosegrasse(C5) anglice rosegras(C6) anglice turmentille and hit is gowd aȝen fenym(C10) tormentille and it is goode ageyn venym(C11) rosegras(C13) yekysters*(C21) columbyne(C21) turmentill(C24) turmentylle(C27) anglice rosegras(C36)
Lit. MS xliiii (antifarmacum); N154

Antitus
Lonicera periclymenum L. Honeysuckle
honysowcle(B4)

Antofili [antafoli, antofoli, autosili, antifolii, antrophilus, atofili, antepholi, anthofali]
Eugenia caryophyllata Thunb. Cloves
gariophe, gylofre, clowe(C9) þe gret clowe yfounde amonge clowys(C10) gelofer(C21) þe grete clow founden among other clowis(C24, C27) anglice grete gelofers(C26)
Lit. A10 (anchofili); A. App. 201 (antophali); MS xlv (antofali); N154

Anton
Rosa L. ssp. Rose
rose-seed(C25)
Lit. see **Antera**

Antos [anthos, antus]
Rosmarinus officinalis L. Rosemary
allipium, rosemary(C21) þe flour of rosemaryne and þe ey withowten scellys(C27) rosemarin(C28)

Antropomos [antopomeros, antopomidros, andropomum, antopomos]
Mandragora officinarum L. Mandrake
mandrake(B3)
mandrake(C2, C12)
Lit. MS xlv (antropomorson)

Aparina
Arctium lappa L. Burdock
Galium aparine L. Cleavers
gallice gletonnere, anglice clot(B1)
clote(C28)

Lit. B III, 52f & 50f. see **Amorfolia**
MS xlvi ('Aparine purpurea, id est lapago')

Apiago
?*Melissa officinalis* L. — ? Balm
Lonicera periclymenum L. — Honeysuckle
honisoke(C6) honysokel(C14)

Apiaria
Borago officinalis L. — Borage
borage(C9)

Apiastrum [apiaster]
Apium graveolens L. — Smallage, Wild Celery
Melissa officinalis L. — Balm
anglice wilde merche(B1)
wodemerch(C25) wodemarch(C28) piment, beowort(C37)

Apium Alexandrinum
Smyrnium olusatrum L. — Alexanders, Horse Parsley
anglice stanmerche(C16) modermerche(C43)

Apium (Commune)
Apium graveolens L. — Wild Celery, Smallage
gallice ache, anglice merche(A1, A4, A6) merche(A3) ache(A8, A11) hache(A9) anglice smalache(A12)
gallice petit ache, anglice smalache vel merch(B1) gallice ache, anglice smalache(B2, B7) ache(B3, B6) smalache, ache, merche(B4) merch(B5) ache, merche(B8)
anglice smaleache (a. domesticum) ache (a. ortolanum) merche(C1) ache, merche(C2) smalache(C3) mersch, smalache(C28)

Apium Agreste
Sanicula europaea L. — Sanicle
Anthriscus sylvestris (L.) Hoffm. — Wild Chervil
Apium graveolens L. — Wild Celery, Smallage
anglice wylde merch or wodemerche or wyld ache or chervylle or syrfoylle(C26)

Apium Amoris
Levisticum officinale Koch — Lovage
anglice loveache(B7)
anglice loveache(C1, C6) lovache(C21)

Apium Emoroidarum
Ranunculus L. ssp. var. Buttercup or Crowfoot
anglice croufote(A2) revenesfo(r)d(A6)
crofot(C1) anglice crakefette(C3) anglice crowfote(C4) rammysfete(C7) sperewort(C26) krowfote(C29) smalache(C35) crowesfote(C39)

Apium Levisticum
Levisticum officinale Koch Lovage
lowache(C16)

Apium Macedonicum
Smyrnium olusatrum L. Alexanders
anglice et gallice staunche [corr. stanmerche] vel alisaundre(B7)
anglice alisaunder(C1) anglice stanmerche vel alisaunder(C6)

Apium Ranarum
Ranunculus sceleratus L. 'Celery-Leaved Crowfoot'
Ranunculus flammula L. Lesser Spearwort
water-crowfot(C1) watermerche(C1) anglice wodrove*(C3) water-crowfote(C4) woderove*(C7) ache dé bois(C17) anglice craufote(C20) watter-crowfoth(C32)
Lit. MS xlvii

Apium Risus
Ranunculus sceleratus L. 'Celery-Leaved Crowfoot'
Levisticum officinale Koch Lovage
Apium graveolens L. Wild Celery, Smallage
?*Anthriscus sylvestris* (L.) Hoffm. ? Wild Chervil
anglice elstong(B1)
cerfoile, chervyle(C1) crowfote(C5) lovach(C7) luvesche(C17) anglice clofunge(C18) smalache(C21) crowtoun(C28) loveache(C35) kerwell, chervell(C42)

Apium Selinum [a. sillinum]
Levisticum officinale Koch Lovage
anglice loveache vel lovesche(C43)

Apium Silvaticum
Sanicula europaea L. Sanicle
Anthriscus sylvestris (L.) Hoffm. Wild Chervil
sanicle(B3) anglice ville(B7)
anglice ville(C5) anglice villen(C6) wodemerch(C7) anglice sanikyl(C12) cerfoile(C21) anglice wod-crowefote(C41)

Apogrissa [apocrifa, appocrissa]
Centaurium erythraea Rafn — Common Centaury
Blackstonia perfoliata (L.) Huds. — Yellow-Wort
cristesladyl [corr. ladir?], seynte Marie ladil [corr. ladir?](C13) the lesse centory(C21)
Lit. André I, p. 36 († apogoris)

Apollinaris [appollinarii, appollinarius, appollinaria]
Convallaria majalis L. — Lily-of-the-Valley
Digitalis purpurea L. — Foxglove
Mandragora officinarum L. — Mandrake
Solanum nigrum L. — 'Black Nightshade'
anglice clofwort(B1) glofwort(B3) glosworte[sic](B4)
(a. maior) anglice foxgloven(C1) glofwort; goldwort secundum quosdam(C2) golwort(C4) golewort(C7) mandrage(C9) closewort [corr. clofewort](C10, C11) anglice glofwort, alibi anglice goldwort(C12) mersmalwe*(C13) (herba a.) foxglove(C14) crepe .i. mavi*(C23) mandrake(C26) clovewort(C28) anglice clofworte(C37) closewort [corr. clofewort](C40)
Lit. B I, 70; II, 53f; III, 114; MS xlviiii ('mandragora')

Apollinaris 2 [appollinarius, apollinaris]
?*Cynoglossum officinale* L. — ? Hound's Tongue
?*Anchusa arvensis* (L.) Bieb. — ? Bugloss
Solanum nigrum L. — 'Black Nightshade'
hondestong(B7)
anglice houndesbery(C6) houndysberye(C13) houndyshert[?](C21) houndbery(C36)
Lit. MS xlviii ('jusqueamus'); B I, 70; II, 70; André I, p. 36

Appollogonus
Polygonum aviculare L. — Knotgrass
unfortrede(C28)
Lit. B III, 241

Appopinatus [apepinatis]
Linum usitatissimum L. — Linseed
anglice lynsede(B7)
lynsed(C13)

Aqua Limphea [for limphea aquatica]
Nuphar Sm. ssp. — Water-Lily

herbe Water*(C22)
Lit. Alph., p.103, 6

Aquilaria
Ranunculus sceleratus L. 'Celery-Leaved Crowfoot'
apium risus apellatur, anglice water-rose(A6)
clofþunge(B3)

Aquilaris [aquilare, aquilaria]
Chelidonium majus L. Greater Celandine
seladyne, malere, ceterwort(B4)
anglice celidon or tetrewort(C4) celyndyne, teterwort(C7) anglice celidoigne, teterwort(C10) teterwort(C12) celidoyne, tetyrwort(C13) tetereworth(C24) celidoyne, tederwort(C27) celendyne, teterwort(C31) celydoyne, setewort*(C40)

Aquileia [aquila, aquilei, aquilaria]
Melissa officinalis L. Balm
Filipendula ulmaria (L.) Maxim. Meadowsweet, Meadsweet, Queen of the Meadow
regina, anglice modwort(B2)
anglice medwort(C5) anglice moderwort(C6) medwort, beerwort, maydewort, reyne de pré, bourfayre(C13) anglice medeworht(C15) moderwort(C16) medewort(C21) reyne, mede-wort(C23) anglice maydenherbe(C41)

Aquileia 2 [aquilona, aquilena, aquilina]
Aquilegia vulgaris L. Columbine
columbyne(B4)
columbine(C10) columbyn(C7, C24) columbine(C27)

Aquileia 3
Sempervivum tectorum L. Houseleek, Sengreen
jovis barba, sengrene(B1)
synegrene(C28)

Aquilina Maior
? ?
argentilla maior, argentille þe more(C1)

Aragia
Atriplex L. ssp. Orache
arage(B5)

Arama
?*Aristolochia rotunda* L. ? Aristolochia, Birthwort
aristologie(C25)

Arangus
Citrus aurantium L. Orange
oreng or appul of oreng(C9)

Aratilla
Allium ursinum L. Ramsons, Wild Garlic
ramese(C28)

Arbor Marie
?*Chrysanthemum vulgare* (L.) Bernh. ? Tansy
anglice fetherfewe(C40)

Arbor Veneris
Umbilicus rupestris (Salisb.) Dandy Pennywort
peniwort(B3)
penyworte(C12)

Arboracia [aboracia, arborancia]
Raphanus raphanistrum L. Wild Radish, White Charlock
Raphanus sativus L./*landra* Moretti ex DC. Wild Radish
wyld radiche(C1) anglice wilde radich(C4) anglice wilde radische(C43)
Lit. cf. **Arborana**

Arborana [aborana, abrona]
Raphanus raphanistrum L. Wild Radish, White Charlock
Raphanus sativus L./*landra* Moretti ex DC. Wild Radish
wylde radisshe(C7) wilde radich(C13) anglice wilde radich(C14)
Lit. cf. **Arboracia**

Arboratica
Polypodium vulgare L. 'Polypody'
ok-verne(C25)
see **Filix Arboratica**

Archalonea
Conium maculatum L. Hemlock, Herb Bennet
gallice herb benet, anglice humblok(B1)

Archangelica [ar(c)ha ang(e)lica]
Lamium album L. White Dead-Nettle
Lamium purpureum L. Red Dead-Nettle
?*Rumex acetosa* L./*acetosella* L. ? Common Sorrel/Sheep's Sorrel
anglice blynd-netyl(A2) morte urtie, blinde-netylle(A3) anglice blynde-netle(A12)
gallice mort urtie, anglice blynd-nettel(B1) treis maner de urtie, anglice blynde-netle(B2) blynde-netle(B3) blynd-netyl(B4) blinde-netll(B5) mort urtyke, blynd-nettyll(B8)
(a. rubea) reed archangele (a. alba) whyt archangele(C1)
blynde-netele(C2) archaungell(C4) archangele, blinde-nettil(C31)
(maior) sourdoke (minor) blynd nettle(C35)

Archapiaris [archaprasis, archapreasis, archaprasis]
Atriplex L. ssp. Orache
melde(C10, C11, C27) melde, brionia, sicida anglice wilde nep(C40)

Archimediocre
Lepidium latifolium L. Dittander
Dictamnus albus L. Bastard-/White Dittany
Origanum dictamnus L. Dittany of Crete
anglice dytayne(C1)

Ardillus [ardilla]
Allium ursinum L. Ramsons, Wild Garlic
anglice rammeson(C10) ramsey(C24) crowegarlyk(C31) anglice romses(C36)
Lit. MS liii ('porrum silvaticum vel vinee')

Ares [arestra, aresta bovis]
Peucedanum officinale L. Hog's Fennel
Melilotus officinalis (L.) Pall. Common Melilot
Ononis repens L./*spinosa* L. Restharrow
wilde trifoyl, hertclaver, camok, ty-hers(C9)

Argentea
?*Lilium* L. ssp. ? Lily
gallice argentine, anglice leele(C16)
Lit. see **Argentine**

Argentilla [augentilla]
?*Filipendula ulmaria* (L.) Maxim. ? Meadowsweet
?*Potentilla anserina* L. ? Silverweed

?*Iris* L. ssp. ? Iris
?*Sedum telephium* L. ? Orpine
(a. minor) argentille þe more, aquelina(C1) anglice orvall(C3) auricula leporina, anglice orvale, orpyn or hariswort(C4) aureum monyle, orval, wyhtgres, goldworth, water betoyne(C9) argentis(C28) stonbrek*(C32) orvale(C36) orvella that knetteþ broken bones togydere(C37) (maior) argentil þe more (minor) argentyl þe lasse(C38)
Lit. cf. André I, p. 39 & II, p. 24

Argentilla 2
Bellis perennis L. Daisy
daysey(C21)

Argentina
?*Alchemilla arvensis* (L.) Scop. ? Parsley Piert
?*Potentilla anserina* L. ? Silverweed
gallice argentile(B1)
Lit. B III, 5

Argentine
Lilium L. ssp. ? Lily
lylye(C25)

Arilli [arulli]
Vitis vinifera L. Grape-Seeds
grapes(B4)
kernels of reysons(C10) kernell of reysons(C11)

Arimel [armiel, armel]
Lithospermum officinale L. Common Gromwell
anglice gromil(B7)
anglice gromyl(C5) anglice gromel(C6) gromel(C21) anglice gromel or whit saxifrage(C26)
Lit. MS liiii ('grana solis')

Arimma [?]
Polygonum hydropiper L. Smartweed, Water-Pepper, Culrage, Arsesmart
erssmert(C28)

Aristolochia [aristologia, astrologia, aristrologia]
Aristolochia L. ssp. Aristolochia, Birthwort
?*Cyperus longus* L. ? Galingale

astriloge(A2) wuderove*(A11)
(rotunda) smereworth(B2) deu manere de astrologie, gallice aristoligie(B2) smerewort(B3) (rotunda) smerwort(B4) smerewort(B5) gallice et anglice aristologe(B7)
(longa) everzwort (rotunda) englich galyngale(C1) (longa) astrology(C4) (rotunda) anglice galyngale, meke(C4) smerworte(C7) (saracine) smerwirt(C13) aristologia la rounde c'est barbote(C17) astriloge(C22) gallice aristologi(C23) smerworth, algea*, floteworth*(C24) anglice long aristologie or red mader*(C26) wodemarch(C28) wodeclote(C28) meke(C29) (longa) red madyr*, emwort[?], long astrolegi (rotunda) astrologi or galyngale(C32) (rotunda) simeringewort vel wodecriate [corr. wodeclate](C37) (rotunda) englysch galyngale(C38) (longa) rede mader* (rotunda) anglice astrologye or galyngale or meke(C41)
Lit. B III, 266

Armala
Peganum harmala L. — Wild Rue
anglice wylde rewe(C26)
Lit. André I, p. 41

Armolitaulus
?*Sinapis arvensis* L. — ? Charlock, Wild Mustard
carlok(C28)

Armoracia [amorceacia, amoreacea, armorica, armoriscia, armoniaca, amorcacia]
Sinapis arvensis L. — Charlock, Wild Mustard
?*Raphanus raphanistrum* L. — ? Wild Radish, White Charlock
anglice carlok(B1) carluk(B3)
.i. cherloke vel werloke vel kerloke(C6) cammok(C7) anglice carlok vel kelk(C12) anglice wylde rape or wyld nept or warlock or charlock; som cal it whit peper or wylde pepir(C26)
Lit. N155

Arnicon [arnica]
Nymphaea L. ssp. — Water-Lily
anglice waterlylye(C26)
Lit. André I, p. 41 (arnion)

Arnoglossa [arnaglossa]
Plantago major L. — Plantain, Waybread
gallice plantayne, anglice weibred(A1) planteyne, weybred(A3) gallice plantayne, anglice weybrede(A4) ribwort(A5) plaunteyne(A7) anglice

weybrede(A9) waybrede(A12)
gallice plauntayn, anglice waybred(B1) gallice planteyne, anglice weybrode(B2) weibrode(B3) waybrede, plantayne(B4) waybred(B6)

Aromatici Anachores [a. anchores]
Acorus calamus L. — Sweet Sedge, Sweet Myrtle, Sweet Flag
Myrica gale L. — Sweet Gale, Bog Myrtle
gallice gaure(B7)
gallice gaure(C5, C8, C36)
Lit. MS lvii (aromaticum, Avic. .i. calamus aromaticus)

Aron
Arum maculatum L. — Cuckoo-Pint, Arum, Adder's Root
calvysfot, wodebane, wild dragance, wodeȝarwe [later hand: wake robyn](C9)

Artasile
Lamium L. ssp. — Dead-Nettle
dede-nettill(C28)

Artemialis
Agrimonia eupatoria L. — Agrimony
anglice egrimoyne(C12)

Artemisia [artemesia, arthemasia]
Artemisia vulgaris L. — Mugwort
gallice armoyse, anglice moderwort vel mogwort(A1) armoyse, modirwrt(A3) gallice armoyse(A4) anglice mugwort(A5) gallice hermoys(A5) mugwert(A11)
anglice vocatur mugwed(B1) gallice ermoyse, anglice mogwed(B2) mogwort(B3) mugworte, ermoyse(B4) mugwort(B6) ermoyse, modyrwort(B8)
mugwort(C1) anglice mogworthe(C3) gallice artemeise vel ermoise, anglice moderwort vel moggewort vel maggwed(C6) (matricaria) hermoyse, anglice mugworth(C10) gallice armoyse, anglice magwed(C18) moggewort(C22)

Artemisia Domestica
Artemisia vulgaris L. — Mugwort
feþerfoie(B5)
anglice feþervoye(C6) febrifuga nostra .i. vetonica, feverfu, anglice

fetherfoye(C7) ferveyne[sic](C13)
Lit. André I, p.134

Artemisia Silvestris
Artemisia vulgaris L. Mugwort
fyuefyver, anglice fetherfoye(C10)
Lit. André I, p. 42

Artetica Maior
?*Plantago coronopus* L. ? Buck's-Horn Plantain
herb yve(C35)

Artetica Minor [herba artetica]
?*Plantago coronopus* L. ? Buck's-Horn Plantain
herbe yve(C14) anglice herbe yve, the beste herbe alyve(C26)

Arthesia
?*Plantago coronopus* L. ? Buck's-Horn Plantain
herbe yve(C21)

Artillus
?*Asplenium adiantum-nigrum* L./ ? 'Black Spleenwort'/
trichomanes L. 'Maidenhair Spleenwort'
saxifraga .i. papwort(C35)

Artitheca [artetica, arterica]
Raphanus raphanistrum L. Wild Radish, White Charlock
Raphanus sativus L./*landra* Wild Radish
Moretti ex DC.
anglice walde rydyche(C3) wilde radik(C23) kerlokk(C29)
Lit. Cf. MS lvii ('primula veris et est herba sicut consolida minor nisi quod habet folia quasi nigra')

Arundo [arunda, arondo]
Phragmites communis Trin. Reed
anglice wyld reed(C12) anglice rosell(C16) reode(C25) rede(C28) red(C35)

Arus
Tussilago farfara L. Coltsfoot
anglice folefote or pé de poleyn(C4)

Asa Fetida
var. umbelliferae esp. *Ferula foetida* Regel and *F. assa-foetida* L. — Asafoetida
gallice glejole(A5)
asa puante(C17)

Asara [asera, asarum, asserum, asari]
Asarum europaeum L. — Asarabacca, Hazelwort
gallice aser(A5)
anglice spewyngwort(C5, C6, C8) avance savage(C7) hesilwrt, wylde molece(C13) spuyngwort(C21) spewinwurt(C23)

Asarab [for asarabacca?]
?*Asarum europaeum* L. — ? Asarabacca
vulgago, anglice uod(C43)

Asarabacca [asara baccara]
Strictly *Asarum europaeum* L., but var. applied
anglice mougwourt(A6)
azarabak(B5)
anglice cameos or alkanet(C4) pervenke(C17) mogwort(C32) avense(C35)

Asaray
Dipsacus fullonum L. — Teasel
wild tesel(C35)
Lit. Alph., p.15, 11 (asar)

Asarus
?*Asarum europaeum* L. — ? Asarabacca, Hazelwort
asara bacora, cariofilata, sauuger(B4)
anglice hariswort [corr. hasilwort?] or coltsfote*(C26) melde(C28)

Asero
Verbascum thapsus L. — Mullein
gallice fost, anglice moleyn(C16)

Aspaltum
?*Capsella bursa-pastoris* (L.) Medic. — ? Shepherd's Purse
nozebledill(C30)

Asparagus [asparagia]
Asparagus officinalis L. — Asparagus

Anthriscus sylvestris (L.) Hoffm. Cow Parsley, Wild Chervil
sparge(A5)
gallice sparge, anglice wodche[r]velle(B1)
wodefille(C1) wodechervell(C28) wylde fylle(C32) wodefyll(C38)
Lit. B III, 265f.

Aspergus
?
powkchese(C21)

Asperiolus
Rumex acetosa L./*acetosella* L. Sorrel/Sheep's Sorrel
Oxalis acetosella L. Wood Sorrel
anglice surel(C5) anglice surell(C6) anglice sorel(C36)

Asphaltion [aspalteos, aspalticos, aspaltos, aspaltia]
Potentilla reptans L. cinquefoil
fyve-leved gres(B4)
fiwe-lefe(C6) fyve-leved gras(C7) anglice .v. levyd gras or synkefoile(C26) synkfoyle(C27)
Lit. André I, p. 44

Aspikaris
Phyllitis scolopendrium (L.) Newm. Hart's-Tongue Fern
?*Asplenium ceterach* L. ? Spleenwort
hertestonge(B3)

Asplenys
Phyllitis scolopendrium (L.) Newm. Hart's-Tongue Fern
?*Asplenium ceterach* L. ? Spleenwort
hertistonge(C9)
Lit. André I, p. 45

Assoala
Sonchus oleraceus L. Sow-Thistle
souththistell(C21)

Asterion [astericion]
Botrychium lunaria (L.) Sw. Moonwort
?*Lunaria annua* L. ? Honesty
?*Aster* L. ssp. ? Aster
asplenis, lunarie, monewort(C1) anglice lunary(C4) romani vocant esteileie, anglice glewort(C43)

Asterinon [or astermon?; asterion]
Juniperus sabina L. Savine
anglice saveyn (C4, C26)

Astila [= astula regia?]
Papaver L. ssp. Poppy
popy*(C13)
Lit. B III, 188f

Atarpassa [atrapassa]
Sambucus nigra L. Elder
anglice elder-blome(C26)

Athanasia [anatasia]
Chrysanthemum vulgare (L.) Bernh. Tansy
tanesi(B3)
tanesey(C9) tanzay(C14)
Lit. Alph., p.16, 23

Athiantos
Adiantum capillus-veneris L. Maidenhair Fern
waterwortys(C1)
Lit. B II, 124f.

Atonia
Trigonella foenum-graecum L. Fenugreek
gallice et anglice fenugreke(C16)

Atrafaxos [atrafaxon, atrafax, andrafex]
Atriplex L. ssp. Orache
afformache(C1) attriplex, cokkowpyntell(C21)
Lit. MS xlviii ('andrafasis'; atrofases 'atriplices'); André p. 31 (andrafaxis) & p. 46 (atrafaxis)

Atriplex
Atriplex hortensis L. Orache
?*Beta vulgaris* L. ? Beet
gallice arache(A1) arache, melde(A3) arage(A7) anglice melde(A12) (minor) meldes (commune) deu manere de meldes (maior) gallice arage, anglice meldes(B2) arage(B4)
(maior) anglice arage, medlus(C1) arage(C2) mylde(C2) (minor) anglice arache vel medles, secundum quosdam cockuspyntel(C5, C6) gallice orage, anglice meldez(C8) bleta .i. betis(C35) (minor) anglice arage vel

medles vel secundum quosdam kockcupintel(C36) wylde arage(C38) bunwode(C39) anglice wolde [corr. melde](C40) anglice arage or mielde(C41)

Atriplex Agrestis/Agreste
Atriplex L. ssp. Orache
anglice melde(A2) anglice meldene(A6) anglice melde(A8) anglice melde(A9)
anglice melde(B5)
(a.a., minor vel alba) wilde or whyt arage(C1) anglice meldys(C3) melde, arache(C3) anglice wilde arage(C4) anglice mylde(C12)
Lit. N156

Atticus [attitus]
?
honysouke(C7) honysokell(C28)
Lit. A. App. p. 203 ('acitus .i. lappatium')

Aureum Monile [aurum m., a. novile]
?*Iris* L. ssp. ? Iris
fane(A8)
unum stipitem habet bene erectum, florem pulcherimum, folia ut lactuca, sed parve sunt strictiora anglice sanell(B9)
anglice sanne [corr. fanne?](C5, C6, C8, C21) fane(C32) (a.m. pratellum) medeworte(C33) sanne [corr. fanne?](C36) sowþernwode*(C37)
Lit. Agnus castus p. 230

Auricula Leporis/Leporina [arula leporina, aricula lepora]
Sedum telephium L. Orpine
gallice horpyn, anglice halsworth(B1) orpyne, halsworte(B4)
hare-eere(C1) hayswort, horpyn(C7) hareswort, orpyn(C10) halswort, (c)orpyn(C11, C21) hare-ere, halswrt, orpyne(C13) harisworth, orpyn(C24) halswert(C28) hare-ere(C38) halyworte [sic], orpyn(C40)
Lit. cf. B III, 121f.

Auricula Muris
Hieracium pilosella L. Mouse-Ear Hawkweed
anglice mouser(A2) orenal .i. musere(A6) musere(A8)
mousere(B3)
mouse-eere(C1) mousere, gallice pelluet(C2) gallice peluet, anglice moushere(C12)

Auriculare/Auricularis
Hieracium pilosella L. Mouse-Ear Hawkweed
anglice mouser(C5) anglice mousher(C6) peluete, mowsere(C13)

Aurilla
Glechoma hederacea L. Ground Ivy
ere terestre, anglice tonhoue(B1)
Lit. B I, 86

Aurum Valet
?*Sedum telephium* L. ? Orpine
.i. argentilla, anglice orvale vel flebat [sic](C36)

Avancia [avantica, avencia]
Geum urbanum L. Wood Avens, Herb Bennet
gallice pé de levre, anglice harefot(A1) avance(A3) anglice avance(A6) avence(A8)
gallice avence, pé de levre, anglice harefot(B2) avence(B3) haresfote(B4) avence(B5) avence, harfoȝt(B8)
(maior) more avanse, harefot (minor) lassa avanse(C1) gallice avence vel pee de levere, anglice harefot(C6) harefoote, avence(C27) romani vocant eam gariofile vel sanamunde(C43)

Avellana
Corylus avellana L. Hazel, Cob-Nut
gallice noyz de coudre(A1) noys de coudre, hasyl-notte(A3)
anglice smale notes(B2)
anglice walnote*(C6) anglice walsnote*(C8) gallice peti noys, anglice smale notes(C18) hasyllnote(C21) haselnote(C22) anglice smal notte or fulbertys(C26) haselnot or heggenote(C31) haselnote vel filbertes(C36) an akkorn*(C39)

Avena [avenato]
Avena fatua L. Oat
anglice ote(C1) anglice otys(C26)

Ayros Maior [ayron]
Sempervivum tectorum L. Houseleek, Sengreen, Ayron
gallice jubarbe, anglice sengren vel houslek(B1)
houslec(C35)

Babacon [for batrachion?]
Ranunculus lingua L. Great Spearwort

Ranunculus flammula L. Lesser Spearwort
sperwort(B3)
anglice sperwort(C12)

Babilonicus [= Cyperus b.]
Cyperus longus L. Galingale
galinggale(C2) anglice galingale(C12) anglice galyngale(C26)

Bacca Edere [bacce e.]
Hedera L. ssp. Ivy-Berry
gallice baye de ere, anglice of ye yvy(A1) baye de ere, yvi cnappis(A3)
gallice le bai de ere, anglice beries of þe ivy(B2) gallice le bay de ere, anglice beris of ivy(B7)
gallice le bay de ere, anglice berion of yvy(C5) baye de ere, beryen de yvy(C13) anglice ivyberien(C20) yvy bays(C21)

Bacca Lauri [bacce l.]
Laurus nobilis L. Laurel-Berry
gallice baye de lorer .i. frut(A1) baye de lorer, laurberie(A3)
gallice bays de lorere(B1) gallice bay de lorere .i. frut de lorere, anglice beryes of þe lorere(B2) anglice beries of lorer(B7)
gallice bay de lorer, anglice lorerberye(C8) lorer bays(C21)

Baccara [asabaccara, asarabaccara]
Asarum europaeum L. Asarabacca, Hazelwort
spywingwort(C10) spewyngwort(C11) spewyngworth(C24)
spewyngwort(C27)

Bachillum
Vicia faba L. Bean
beene(C13) bene(C21)
Lit. MS lxxii (bakilla 'faba')

Bacina [batina]
?*Mandragora officinarum* L. ? Mandrake
mandrake(C21)

Bacticia
Pastinaca sativa L. Parsnip
pastinake, pasternepe(C10)

Baculus Pastoris
Verbascum thapsus L. Mullein, Shepherd's Staff

Dipsacus pilosus L. — Small Teasel, Shepherd's Rod
schephurdisstaf(C1)

Badga
Cynoglossum officinale L. — Hound's Tongue
hunde-tynde vel -tong(C3)
Lit. André I, p. 52 (bat(t)ica)

Balanon [balanona, balanus]
Quercus L. — Oak-Apple
ok(C9) anglice akorn(C26)
Lit. André I, p. 50 & II, p. 32; Alph., p.18, 29

Balaustia [balustria, balaustica, balaustion, balusta, balstria]
Punica granatum L. — Pomegranate
gallice la flur de pome gernet(A1) la flour de gernette(A3) gallice balaustres(A4) anglice wylde lilie(A5) flour de poumes g[ar]ne[tes](A7) gallice le frut de gernetes, anglice þe flour of gernete(r)(B2) gallice le frute de gernete(B7)
flowre de pome garnett(C3) le fruyt de granet(C6) the whiȝt lyly*(C21)
Lit. N156

Baldemonia [baldimonia]
Gentiana L. ssp. esp. *amarella* (L.) Börner — Gentian/Felwort
?*Meum athamanticum* Jacq. — ? Spignel, Baldmoney
baldemoyn(B1)
genciane(C16) baldemoyn(C28)

Ballota
Ballota nigra L. ssp. *foetida* Hayek — Black Horehound
?*Marrubium vulgare* L. — ? White Horehound
anglice whyt watermynt* or horhound(C26)
Lit. André I, p. 50 (ballote)

Balsamita [balsamenta]
Mentha longifolia (L.) Huds. — Horse-Mint
Mentha aquatica L. — 'Water Mint'
gallice balsamite, anglice batym .i. horsmint(A1) watermentes(A3) anglice horsmente(A6) horseminte(A8) anglice horsmente(A12) brome*(B1) gallice balaunte, iii maner de mente, anglice batin, water[minte](B2) horsmynte(B3) horsemynte, mentastere(B4) anglice horsemente(B5)

anglice watermynte(C3) gallice balsami(n)te, anglice waterminte vel horsmynt(C5) gallice balsamite, anglice waturmynte vel horsemynte(C6) balsamy(n)te, watermynte, horsmynt(C13) basm, water[mynte](C23) whit watermynt or whit bame(C26) brounwort*(C28) herbe water*, horsmynte(C37) (rubia) red horsmynte(C38) (rubia) anglice horsmynte or spryngwort or watermynte(C41) horsemynt, medemynt, watermint(C42)
Lit. Alph., p.19, 17

Balsamum [balsamus]
Gum of *Commiphora opobalsamum* Engl. — Balsam Tree
gumme de baumer, holy wey(A3)
gallice baume e la gumme de baume, anglice hali wey(B2) gallice baumer, anglice hali wei(B7)
gallice baumier, anglice halwey(C6) gallice baume, anglice haly whey(C8) holowberys* [sic](C22) hali wey(C28)

Balsamus
Commiphora opobalsamum Engl. — Balsam Tree, Balsam of Gilead
baumer(A3)
gallice baumere(B2)
bawme-tre(C2, C21)
Lit. N157

Banados [banodes, banades]
Sempervivum tectorum L. — Houseleek, Sengreen
synegrene vel houslek(C8) anglice singrene vel houslek(C36)

Banarella
?*Capsella bursa-pastoris* (L.) Medic. — ? Shepherd's Purse
?*Polygonum aviculare* L. — ? Knotgrass
anglice blodwort(C36)

Bantica
?*Cynoglossum officinale* L. — ? Hound's-Tongue
anglice houndestonge vel buglossa*, romani vocant chenlange(C43)
Lit. André I, p. 52 (bat(t)ica)

Barascus
?*Atriplex* L. ssp. — ? Orache, Milds
?*Chenopodium album* L. — ? Fat Hen

anglice middewort, gallice melie [corrected in MS from medie] .i. me[l]de(B1)

Barba Aaron [A(a)ron b.]
Arum maculatum L. Cuckoo-Pint
anglice yekesters(A1, A3, A4) jare, yekesters(A3)
gallice sacherey*, anglice gotis-pintil*(B1) gallice jare, anglice ȝekpintel(B2) gallice jare, anglice cokousepintul vel ȝekestres(B7)
cokkowespintel(C2) anglice calefote(C3) gallice jare, anglice cokkyspyntel or rampe vel ȝekesters(C6) pee de pulayn, ȝekesters(C10) c'est(e) barberole(C17) ȝextere(C24) anglice cowort [corr. cocuuort?] or calsfote or cowkepyntel(C26) cucubespintel [sic](C28)
cockys-floure(C33)
Lit. N146

Barba Angelica
Lamium album L. White Dead-Nettle, Archangel
modirworte*(B4)
Lit. cf. Alph., p. 20, 10

Barba Elixis
Arctium lappa L. Burdock
clote(B3)
clote(C2)
Lit. Alph., p. 20, 17

Barba Jovis
Sempervivum tectorum L. Houseleek, Sengreen
jubarbe, selfgrene(A3) anglice stoncrop*, gallice jubarbe, anglice silvegrene(A5) anglice sifgrene(A6) anglice huslek vel singrene, gallice jubarbe(A8)
gallice jubarbe, anglice sindgrene vel houslek(B2) gallice jubarbe, anglice senegrene(B5)
howseleke(C3) jubarbe, horewort [for erewort?](C7) erewert, syngrene, houselek(C13) jubarbe, selfgrene(C22) selgrene(C27) hous-lily [sic](C28) frenche rewbarbe*, anglice rewbarbe*(C34)

Barba Leporis
Sonchus oleraceus L. Sow-Thistle
Verbascum thapsus L. Mullein, 'Hare's Beard'
thowthistyll(C3) rosa passa, anglice hareberd(C12) harebell*(C29)
Lit. Alph., p.154, 39 (rosa passa, barba leporis idem)

Barba Salomonis
?*Plantago coronopus* L. — ? Buck's-Horn Plantain
anglice herbe yve(C36)

Barba Silvana
?*Sempervivum tectorum* L. — ? Houseleek, Sengreen
?*Alisma plantago-aquatica* L. — ? Water-Plantain
housleke(C21) plantago, anglice weybred(C39)

Barbastus [for thapsus b.]
Verbascum thapsus L. — Mullein
moleyne(C2) anglice molayn(C3)

Bardana [bardona]
Arctium lappa L. — Burdock
Galium aparine L./*palustre* L. — Cleavers/'Marsh Bedstraw'
anglice clote(A2, A12) bardane, glettenere, clote(A3) anglice clyte(A5) gallice gletinere, anglice clote(A6) gletounere, clote(A7)
anglice papewort*(B1) gallice bardaner vel gletoner, anglice clote vel boure(B2) clete(B4) anglice clote(B5)
anglice burdok(C3) gallice bardane vel gletener, anglice clote vel bur(C6) (magna) waterbardan, waterclete(C9) gletineyr, clete(C15) gallice bardon vel glitoner, anglice clote vel burre vel schalgras(C16) la grant lappe(C17) cloter(C27) glutunere, anglice clote(C43)
Lit. N157

Barrago
Ajuga reptans L. — Common Bugle
Ajuga chamaepitys (L.) Schreb. — Yellow Bugle
wodebroune(C13) bugle(C21, C36) bugile(C28)

Basilica [basilicus, basilisca]
Polygonum bistorta L. — Snake-Root, 'Bistort'
Dracunculus vulgaris Schott — Dragon Arum
gallice dragance ou serpentarie, anglice neddirwort(A1) serpentinerie, edderwrt(A3)
gallice serpentarie, dragance, anglice neddrewort(B2) naddrewort(B3) dragance(B3) anglice neddurwort(B7)
dragaunce(C2) gallice serpentari vel dragaunce, anglice neddyrwort(C6) dragaunce, addyrwort(C13) nedergresse(C14) addirwort or dragance(C21) serpentyne(C22) nedirwort(C28) hedderwort(C35)
Lit. N157f.

Basilicon
Polygonum bistorta L. — Snake-Root, 'Bistort'
?*Dracunculus vulgaris* Schott — ? Dragon Arum
colu[m]bine(C28) þe lasse basilicon(C38)

Basilisca [basilica]
Gentiana L. ssp. esp. *amarella* (L.) Börner — Gentian/Felwort
feldwert(A11)
.i. gentiana, anglice baldemoyn(B7)
feldewort(C28)
Lit. Alph., App. p. 205

Batracion [butracium, brutacion]
Ranunculus sceleratus L. — 'Celery-Leaved Crowfoot'
Ranunculus acris L./*bulbosus* L. — 'Meadow Buttercup'/'Bulbous Buttercup'

anglice cloftong(B1)
clofetunge(C28)
Lit. B III, 53f

Batus [baccus]
Rubus fruticosus agg. — Blackberry
Morus nigra L. — Mulberry
brembell(B6)
anglice rede brere(C3) anglice brerys or blembris, blackberyes(C26)
(rubeus) anglice mulleberyes(C34)

Baucia [bauciana]
Pastinaca sativa L. — Parsnip
Sium sisarum L. — Skirret
?*Anthriscus sylvestris* (L.) Hoffm. — ? Cow Parsley
Heracleum sphondylium L. — Cow Parsnip
gallice pastinak, anglice pasternepe(A1) pastinac(A3)
gallice pastinak, anglice wilde skirezwit(B2) scirewyth(B6) anglice pasnepe vel wylde scirwhit(B7)
(maior, ortulana, domestica) more skyrwhyt(C1) passnepys(C3) anglice pasturnep vel wylde skyrw[it](C6) gaser(C6) pastinake, pastrenep, pasnet(C11) pastnep, wilde skirwhit(C13) pastnake, pastnep(C24) anglice wyld pasnept(C26) pastinake, pastrnepe(C27) cow-wort(C31)
Lit. N158

Bedegar [bedagar, boldegar, bedagord, bedogar, bedigar, bedagrage]
Rosa rubiginosa L. Eglantine
Rose canina L. Dog Rose
Crataegus monogyna Jacq. Hawthorn
?*Trapa natans* L. ? Water Caltrop
?*Eryngium maritimum* L. ? Sea Holly
gallice englenter(A2) eglentere dé boys(A3) anglice wyte breres(A6) anglice heþezorn(A10) vulgare nostro engleter sive botuner(A10) gallice eglenter dé bois, anglice hepe-tre(B2) eglenter, croppes of brere(B3) bremel, eglentire(B4) hepe-tre(B5) gallice orvale*(B9) eglentere or hebrembel(C2) qwyte rose(C3) gallice eglentere dé boys, anglice hepe-tre(C6) eglentyn, hep-brer(C7) zeobrer(C10) ȝhebrer(C11) eglenter, heþeþorn(C13) casbery(C14) anglice whithorne(C16) anglice heopebler [sic], gallice englenter(C19) anglice hepesberie(C20) whight rose(C21) hupbrere(C21) wylde rose(C23) ȝhebreer, eglenter(C24) anglice eglentyne, hoppebrere, swete-brere(C26) eglenter, ȝelwberye(C27) eglentyne(C34) wytþorne(C35) qwythorn(C39) zeobrere(C40)

Been [beenus, beena]
Moringa Burm. ssp. Behen also var. applied to nut-trees and others
bem, ben, notebem, quykebem, whitel-tre, foseyn(C9)
Lit. N158

Bellerici
Terminalia bellerica L. Belleric, Bastard Myrobalan
gallice belrik, un espece de m[ir]abolanes(B2) gallice bellerik(B7) gallice bellerik(C6) gallice bilryk, un espice de mirabolon(C16) gallice bilrik, un espece de mirabolanes(C18)

Belsoguasen [= arab. Barsijauschan; belsognason, bestonguasten, besceguascen]
Adiantum capillus-veneris L. Maidenhair Fern
anglice maydenher(C8)

Benedicta
Conium maculatum L. Hemlock, Herb Bennet
Cicuta virosa L. Cowbane
Geum urbanum L. Wood Avens. 'Herb Bennet'
benet(C1) herb benet, hommelok(C10) benet, homloke(C11) herb benet,

hemelok, wodewistle(C13) herb benet, homlok(C27) benet, hares-here(C32)

Bengli [bongli, bengi vel bengli, bengana]
Hyoscyamus niger L. Henbane
hennebane(C8, C21) anglice henbane(C6, C26)

Berberis [beberis]
Berberis vulgaris L. Common Barberry
un maner de frut(B2) anglice barberyus(B7)
anglice barbareuys(C5) anglice bereberiis(C6)

Bersiendarius [beribendarius, borciendarius, bersiendarii, barsiendarii, bersiendaris]
Dipsacus sylvestris Huds. Wild Teasel
anglice wylde tesyl(C6) wild tasil(C13) anglice wyld tesyll(C26)

Berula [burrula, birula, berilla, berbula]
?*Veronica beccabunga* L. ? Brooklime
?*Berula erecta* (Huds.) Coville ? 'Narrow-Leaved Water-Parsnip'
?*Rorippa nasturtium-aquaticum* (L.) Hayek ? Watercress
anglice bilerne(A2, A12) gallice [sic] bulerne(A6) anglice bilrin(A8) anglice billerne(A9)
bilren(B3) anglice bilre(B5) anglice billeren(B9)
(b. maior, aquatica) more billurs (b. minor, agrestis) lasse billurs(C1) billeres(C2) anglice billers(C3) gallice ber[le], anglice bylren(C6) bere, bilren, berle(C13) byllern(C21) birle(C24) anglice waterlemke or byldryse(C26) anglice burle(C39) bellyre(C42)
Lit. B III, 27f., Alph., p. 21, 21.

Beruyle
? ?
vervayne(B4)

Beta
Beta vulgaris L. Beet
anglice betis(B1) wild nepe(B6)
(b., bleta alba) whyte betys (b., bleta rubea) red betys(C1) betes(C2) anglice betes(C3) gallice et anglice betys(C16)

Betagensis [betegensis, begetensis]
Artemisia abrotanum L. Southernwood
gallice averoyne, anglice suthernewode(C8) sothernwode(C21) sowtherwode(C36)

Betonica [vetonica]
Betonica officinalis L. Betony
gallice betoyne, anglice betonk(A1) betoyne(A3) gallice betoyne(A4) betoyne(B1) gallice et anglice betoyne(B2) betoyne(B3) betayne(B4) botoyne(B5) betumeke(B6)
bytayne(C1) gallice betoyne, anglice buschoppuswort(C6) bittayne(C7) gallice et anglice betayne(C16) beteyn(C21) betayn(C28) gloswort*(C37) woderove haþ rede floure(C37) betonie or brone surwort(C43)

Bibos
Foeniculum vulgare Mill. Fennel
fenkel(C21)

Bifrage
Nuphar lutea (L.) Sm. Yellow Water-Lily
Nymphaea alba L. ssp. alba White Water-Lily
edokke(B6)

Bifuscus
?*Veronica beccabunga* L. ? Brooklime
?*Berula erecta* (Huds.) Coville ? 'Narrow-Leaved Water-Parsnip'
?*Rorippa nasturtium-aquaticum* (L.) Hayek ? Watercress
gallice bere, anglice bilre(C23)

Bilaferos
?*Arctium lappa* L. ? Burdock
clote(C37)

Bironia [for brionia?]
?*Verbascum thapsus* L. ? Mullein
?*Bryonia dioica* Jacq. ? Bryony
anglice tapirwurt(C23)
Lit. C13 = 'brionia . . . flosmus . . .'

Bisaco [bisaca, bysaga for besasa]
?*Peganum harmala* L. — ? Wild Rue
?*Thalictrum flavum* L. — ? 'Common Meadow Rue'
violett*(C9)
Lit. André I, p. 53 (bessasa)

Bisara [bysarya, bysoria, bisasa]
?*Peganum harmala* L. — ? Wild Rue
?*Thalictrum flavum* L. — ? 'Common Meadow Rue'
pety rue, wilde rue, wode rue, woderove*(C9)
Lit. André I, p. 53 (bessasa)

Bismalva
Althaea officinalis L. — Marsh Mallow
wymalve vel mersmalwe(A2) wyte malve vel holyhocke(A6) gallice wimauve(A8) wilde malwe(A12)
gallice wymave(B5) holy hokke(B8) anglice wit malue(B9)
bimalve(C1) anglice wymawe(C3) anglice holihocke(C20)
see-hokke(C31) wilde malowe(C39)

Bistoria
?*Berula erecta* (Huds.) Coville — ? 'Narrow-Leaved Water-Parsnip'
?*Rorippa nasturtium-aquaticum* (L.) Hayek — ? Watercress
?*Veronica beccabunga* L. — ? Brooklime
anglice byllers(C3)

Bistorta
Polygonum aviculare L. — Knotgrass
Polygonum bistorta L. — Snake-Root, 'Bistort'
?*Potentilla erecta* (L.) Räusch. — ? Common Tormentil
anglice harisgras(B1)
genicle, pety bryddestonge, pety porcelan, swynesgres, þe myd bistord(C9) hethnote, tormentille(C13) pervynk*(C21)
Lit. Alph., p.147, 12

Bix
Aconitum anglicum Stapf — 'Monkshood'
napellus, anglice male herb(C36)

Blaverole [blaverele, blaverel, baverole]
Centaurea cyanus L. — Cornflower, Bluebottle

?*Campanula rotundifolia* L. Harebell
gallice rubeseym, anglice blowed(B1) anglice bledewort(B2) blowode(C2) blowort(C5, C8) blowart(C6) anglice b(l)edewort(C16) anglice bledewort(C18) bloowort(C21) anglice blowurt(C23) blowort(C36)
Lit. Cotgrave 'Blaveoles: Blew bottles, Blew blawes, Corne flowers'

Blecatica
?*Anchusa arvensis* (L.) Bieb. ? Bugloss
lang de boof(C21)

Bleta [bletum, blita]
Beta vulgaris L. Beet
betes(A3) bete(A7)
[for clita?] gletuner*, anglice clot*(B1) gallice et anglice betes(B2) (b. alba) why3t betys (b. rubea) red betys(C1) (alba) white blete (rubia) red blete(C38)
Lit. B III, 50f; André I, p. 55

Blitis
var. Chenopodiacae
gallice glitiner*, anglice clite*(A2)
Lit. B III, 50f

Bolum Mushroom
gallice musserun(A5)

Bombix
?*Carum carvi* L. ? Caraway
?*Cuminum cyminum* L. ? Cumin
cardumene, anglice budthistil(C20) cardumena id., anglice bunthistil(C20)
Lit. cf. André I, p. 56 (bombycias) see **Cordimena**

Bonbax
?*Beta vulgaris* L. ? Beet
bete herb(C21)

Bonefaciaria [bonefacia]
?*Geum urbanum* L. ? Wood Avens
avence(C23)

Bongrossa
?*Brassica rapa* L. ? Turnip
rede neppe(C37)

Borago [boraca]
Borago officinalis L. Borage
Anchusa arvensis (L.) Bieb. Bugloss, 'Wild Borage'
gallice borage(A1) borage(A3)
gallice et anglice borage(B1) anglice borage(B2) borage(B6)
(maior) borage (domestica) þe more borage (agrestis minor) wild borage, galwort(C1) (silvestris) wilde borage or calwort(C13) borasche(C37) (silvestris) wylde borage or katwortte [corr. kalwortte?](C41)
Lit. Agnus Castus p. 213

Boray
? ?
blienicus, taywede(C21)

Botor Marien [= arab. Buchur marjam?; bucor marine]
Cyclamen hederifolium Ait. Sowbread
erȝnothe(C1)
Lit. B III, 84f.

Branca Ursina
Acanthus mollis L. Bear's Breech
gallice la grant mave*, anglice holihok*(A1) anglice sedocke(A2) la grant mauve*, holyhokke*(A3) anglice seddocke, wolvedocke(A6) anglice sedocke(A9) sedok(A12)
brank-ursine(B1) gallice la grant mave*, anglice holihok*(B2) sedokke(B3) anglice se[d]ucche vel malve(B5) anglice holihok*(B7) anglice sedok(B9)
seedokke(C1) anglice sowredoke*(C3) beris-cheke, quidam dicunt sedok(C7) berewort, berefot, branc-ursyn(C9) branc de ursine(C10) braunche-ursyn(C11) gallice branc ursine, anglice seedok(C12) brank-urcine or cutbertull or seedokke(C13) gallice grant malve*, anglice holyhoke*(C16) anglice sedech vel merskmalve(C20) brank-ursine(C21) la grant mauve*, anglice holyhocke*(C22) branch-ursine(C27) see-holy*, or water lyly*(C29) brankursy(C31) cutbertol, sedokke, brank ursyn, berfoot(C32) hocckis*(C35) berefot(C36)

Branci [bransi, baruci, brancia, brancea]
Sium sisarum L. Skirret
Pastinaca sativa L. Parsnip

Pastinaca sativa L. Parsnip
gallice pastinape, anglice siler vel wyld skyrwyt(C5) gallice pastinap, anglice syler vel wylde skyrywt(C6) gallice siler, anglice wilde skyrewit(C8) anglice pastinepe(C12) anglice wyld skirwyt(C16) skyrwete(C21)

Brassica [brasea, brascia]
Brassica oleracea L. Wild Cabbage
gallice cholet savage, anglice wyld coul(A1) cholet savage, wylde coul(A3) wylde coule(A3) anglice red cowele(A5)
anglice cawel(B1) gallice cholet savage, anglice wilde coles(B2) chelet [corr. cholet], wortes(B4)
gallice cholet savage, anglice wilde cole(C5) caulys nondum transplantata, unsetteworte(C9) wilde coole(C13) chol qui n'ad pas esté plantee(C17) gallice cholette savage, wilde coul(C18) colet, caule-wortis(C24) coolwortys(C27) red col(C35) feld caule(C37)

Brassica Silvatica
Brassica oleracea L. Wild Cabbage
Rumex L. spp. Sorrel
anglice wodesure; alio modo dicitur caulis non transplantatus vel quilibet caulis(B1)
wodesure(C28)

Brathios [bracteos, brakseos]
Juniperus sabina L. Savine
savyne(B4)
anglice savyn(C3) savyn(C24) anglice savayne(C39)
Lit. André I, p. 57 (brathy)

Brigla
? ? 'Cripplewort'
anglice cripulwort(B7)
anglice croppeleswort(C5) anglice crepleswort(C6) cropeswort(C8) crepilwirt(C13) croplesewort(C19) crepeliswort(C21) cropleswort(C36)

Brionia [briona, blionia]
Bryonia dioica Jacq. White or Red Bryony, Wild Nep
anglice wild nepe(A2) brionye, wylde nepe(A3) gallice brioyn(A5) anglice wilde ney(A6) wilde nep(A7) wilde nepe(A8) anglice wilde nep(A9) uode nepe(A10) anglice wilde neip(A12)
gallice brioyne, anglice wilde nepe(B1) gallice brionie, anglice wilde nep(B2) wilde nep(B3) wilde nep(B5) anglice wilde nep(B5)

anglice wild [n]ep vel taperwort(C8) brian(C10) anglice taperwort(C12) caperworth, wodenepe(C13) brioyne, wylde nepe(C22) anglice wylde rape or wyld nept or wodenept or taperwort(C26) lynge(C27) wylde nep, bryan(C31)

Britannica
?*Primula veris* L. ? Cowslip
anglice primerole(B1) hawenehydele(B3)
primerose(C28)
Lit. B III, 127 (meaning unknown); cf. André I, p. 58 & II p. 38

Bruerium [bruerum, bruere, bruerie]
Calluna vulgaris L. Ling, Heath, Heather
heþe(C1, C38) hethe(C6) hoth, heth(C13)

Bruncus [brunca, broncus]
Cuscuta europaea L. 'Large Dodder'
Cuscuta epithymum (L.) L. 'Common Dodder'
?*Eupatorium cannabinum* L. ? Hemp Agrimony
doder(C7) dodyr(C9) dodur(C24) anglice dodyr or wyld hymp(C26)

Bruscus [brittus]
Ruscus aculeatus L. Butcher's Broom, Knee Holly
anglice bonwort(A2) anglice woxesglove*(A6) bonewort(A12) gallice haynere, anglice holu vel holyn(B1) gallice flagroun, anglice cnouholin(B2) bonwort(B3) countewort vel wylde box(B7) anglice brusk(B9)
bonwort vel knowholin(C2) brusche(C3) gallice fragun, anglice licholyne(C5) anglice contewort vel wylde box vel knocgalyn(C6) gallice fragun, anglice holyhocke*(C6) bonworte(C7) knochelyn(C7) countwort vel wilde box(C8) bonewort or kneholm(C10) vnobholein[sic], bruske(C10) bnobholm[sic], brucke(C11) gallice fragun vel fragunce, anglice bonwort vel knowholin(C12) fragun, kneholin, wylde box, countewrt(C13) gallice flagroun, anglice knouholin(C18) anglice bunwort(C20) anglice lichelme(C21) wild box(C21) knouholm, brusk, bonwort(C27) holyne(C28) wodebynd* vel ragwed*(C32) gallice fragun, anglice lycholme(C36) bowewort [corr. bonewort?](C37) benworth [sic](C39) fragun, anglice bruscks(C40)

Bubul
?*Origanum dictamnus* L. ? Dittany of Crete
?*Dictamnus albus* L. ? White Dittany

ditayne(C1)

Bugla [bugula]
Ajuga reptans L. Bugle
often conf. w. Buglossa
gallice bugle, anglice wodebroune(B2) wodbrome, bugile(B4) bugle(B5)
(b. Alexandri) bugil Alisaunder (b. maior) whyt bugyl (b. minor, nigra) brown bugil, wodebrowne(C1) bugile(C2) anglice oxtonge*(C3) bugle, celfhele, bugle þe more and þe lasse(C9) bugle or selfhele(C26) bugle, wodeborun [sic](C27) bugil, wodebroune(C31) broun begel(C32) (b. Alexandri) white bugyll (minor) brown bugyll vel wodebrown(C38) browne bugill, silfehele, hartwort(C42)
Lit. Agnus Castus p. 224

Buglossa [buglosa, buclossa, bucosa]
Anchusa arvensis (L.) Bieb. Bugloss
often conf. w. Bugla
anglice wodeburne .s. bugle*(A2) bugle*, ruþerestunge(A3)
gallice bugle, anglice foxeglove* vel selfale(B1) gallice lange de boef, anglice reþerne-tounge(B2) oxetonge(B3) oxetonge, longe de bef, rostus de buse [= rest-beof?](B4) ruþerstunge(B5) bugle*(B6) bugle*, wodbroune*(B8)
oxetonge or netystonge(C9) langydbyf or long de bef(C11) gallice langebef, anglice oxetonge(C20) lang de boof(C21) barba silvana, lang de bef(C21) retherestonge(C22) anglice retheristonge(C23) wodebrome*(C28) gallice longe de befe, anglice oxtung or brownwort* [= bugle](C41)
Lit. B III, 98f; N160

Bulbus
Allium cepa L. Onion
red oynoun(B3)
(communis) anglice a red onyon(C12) anglice oynons or chyboldes(C26) feld lily(C28)

Bulbus Scilliticus [bulbi scilitica, b. squilitius]
Iris L. ssp. Gladden
Gladiolus L. ssp. Gladiolus
batracion*, hoc est gladene(A11)
gladene(B3)
gladene(C2, C28)
Lit. B III, 113; André I, p. 61 & II, p. 40
Bulbus Vomitorius

Bulbus Vomitorius
Iris L. Gladden
gladene(B3)
gladene(C12)
Lit. André I, p. 61 & II, p. 41

Bulmago [busmago]
Ononis repens L. Restharrow
anglice mockys or uynnys or cammok(C26) gallice restebef, anglice cammok or yserneherde(C41)
Lit. Alph., 25, 5

Buphthalmus [bultalmus, butalmon(e)]
Chrysanthemum leucanthemum L. Ox-Eye Daisy
Primula veris L. Cowslip
anglice malve*(B1) cowsloppes, oyle de bef(B4)
cowsleppe(C2) oyle de bef, couslop(C24) anglice daysye or bryswort(C26) oyle de beif, kousloppe(C27)

Burgh
Dipsacus fullonum L. Teasel
anglice tosul(B7)

Burit [buritus, burik, barit, bariche, buroth]
Saponaria officinalis L. Soapwort
saponarie, crauwesope(A3)
gallice saponarie, anglice crousope(B2)
anglice tesyll vel crowsope(C6) savener(C10) gallice saponere, anglice crowsop(C16) tasyll(C21) crowsope, savonner(C27) crowsope, browort(C31) tesel(C36)
Lit. N159

Burneta [burnetta, bruneta, burnetum]
Sanguisorba officinalis L. Great Burnet
Pimpinella saxifraga L. Lesser Burnet
Poterium sanguisorba L. Salad Burnet
burnete(A3)
gallice et anglice burnet(B2)
(maior, agrestis) þe more burnete (minor, pratina) þe lasse burnete(C1) burnete(C2) anglice burnett(C3) gallice et anglice burnett, sprungwort(C16) burnet herb(C21) anglice burnet or spignel or medewort(C26) bernete(C32) (pratina) medwe-burnete(C32)

Bursa Pastoris
Capsella bursa-pastoris (L.) Medic. Shepherd's Purse, 'Sanguinary'
Polygonum aviculare L. Knotgrass
burs de pastur, suinescarse(A3) gallice burce de pastour, anglice sanguinarie, stanche(A3)
gallice bource de pastour vel staunche vel sanguinarie(B2) hencarses(B3) hen-cres(B4) sanguinaria .i. aumener, anglice swnescarse(B5) (maior) bischopyswort (minor) hennec[re]sse(C1) anglice staunchblode(C5) anglice stancheblode(C6) anglice staunchеblode vel swyneskersys(C6) anglice hennekersed(C12) shepperdespurse vel toþwort(C13) staunches(C14) estanche(C17) stanche vel sanguinarie(C18) stauncheblode(C21) burse de pastor, swynyscarse . . . stanche(C22) staunche(C27) caswede(C28) hencres(C29) blodewort, swynisgras(C31) purswort(C34) sanguinaria minor, .i. cassewed(C35) stanchblod vel sanguinaria vel hencres(C36) pursewort(C37) byschopiswort(C38) shepherdpurs or caswede or tothewort(C41) anglice foreglove* [corr. foxglove*] vel henne-cressen(C43)

Butramnus
Rosa canina L. Dog Rose
Rosa rubiginosa L. Eglantine
butenes, hep(C28)

Cabus
Brassica L. ssp. Cabbage
anglice whit caule or chabochys(C26)

Cacabulum
Solanum nigrum L. Black Nightshade, Petty Morel
litil morell(C21)
Lit. André I, p. 63 & II, p. 43

Cachabelma [cantabilma]
Allium vineale L. Crow Garlic
anglice crouelek(B1)
crowleke(C28)

Caculus [cacula]
Daucus carota L. Wild Carrot
?*Athamanta cretensis* L. 'Cretan Carrot'
dauk, carvi,(C9)
Lit. Alph., p. 30, 17 & DMLBS (careum)

?Mercurialis perennis L. ? Dog's Mercury
Chenopodium bonus-henricus L. All-good, Mercury, Smearwort
mercuri, papwort, more smerwort(C21)

Calamentum Maius
Calamintha L. ssp. Calamint
gallice calamente(A1) calamente(A3)
une manere de calamente(B2) calament(B6)
(c. domesticum, ortolanum) calamynte (c. agreste) wyld calamynte(C1)
anglice calamynte(C3) anglice nepte vel calamynte(C6)

Calamentum Minus
Nepeta cataria L. Catmint, Catnip
gallice nepte, anglice cateminte(A1) nepte, kattismete [corr. -mente?](A3)
gallice nepte, anglice cattesmynte(B2)
cattesmynte(C2) (c. domesticum) anglice nept or cattismynt(C26)

Calamistrum
?Galium aparine L. ? Cleavers
cliver(C28)

Calamus
Reed incl. Sugar Cane (Saccharum officinarum)
(c. maior) a reed cane (c. minor) reed cane (c. zuccura) sugircane(C1)
rede(C21) þe swote reede(C31)

Calcesta [calcista]
Trifolium repens L. White Clover
[t]refle, anglice wit clavere(B1)
white claver(C28)
Lit. André I, p. 65; B III, 51f.

Calcipes [calchiepes]
var. caltrops
gallice kaltrappe(C19)

Calendula
Calendula officinalis L. Pot Marigold
rodwort(A3)
gallice solsicle, anglice rede flour vel golde(B2) rode(B3) marigoulde, solsecle(B4)
goldwort or roden(C2) anglice rode flowre vel seynt Marie goldys(C6)

goldwort or roden(C2) anglice rode flowre vel seynt Marie goldys(C6) solsecle, goldewort, ruddis, holy gold(C7) goolde, rodene(C10) seynt Marye rude(C11) ligustrum, primerole*(C14) gallice solsequie vel schekil[sic], anglice golde or rod flowre(C16) rodewort(C22) rodone(C24) gold-kerse(C35)

Calopodia
Tussilago farfara L. Coltsfoot
pee de chivale, anglice horshove(B1)
horshove(C28)
Lit. André I, p. 63 (caballopedia)

Calta
Trifolium pratense L. Red Clover
red claver(C28)

Caltha Silvatica
Trifolium L. ssp. Clover
anglice wodeclavere(B1)
wodeclaver(C28)
Lit. B III, 266

Camabaria
?*Asperula odorata* L. ? Woodruff
corona regia, woderove(B3)
hasta regia, herba muscata, wodrove(C12)

Camamilla [?error]
?*Equisetum* L. ssp. ? Horsetail
cauda pulli, anglice cattepipes(C20) smerewort(C43)

Camatreos
Teucrium chamaedrys L. 'Wall Germander'
burnett*(B6)
Lit. André I, p. 84 (chamaedrys) & II, p. 59

Camce Luvestica
Trifolium L. ssp. esp. *repens* L. Clover, White Clover
anglice witte claver(B1)

Cameactus [cameatricis, cametricis, cameas, camessate]
Sambucus ebulus L. Danewort, Dwarf Elder
eble, walwort(A3)

eble(C2) walwort(C6) ebbyle, walwort(C11) anglice eble(C12) eble, walwort, hareglove(C13) anglice liþewort(C19) anglice walwort or blodewort(C26) hellenewort(C28)
Lit. André I, p. 83 (chameacte)

Camealta [for Camellea = Chamaeleon]
Dipsacus sylvestris Huds. Wild Teasel
wolvescomb(B3)
Lit. B III, 268f.

Camedaphne [camadafnes, camadafnel]
?*Potentilla reptans* L. ? Cinquefoil
?*Ranunculus* L. ssp. ? Crowfoot
grant morele* vel burnette*(A11)
pé de corf, anglice ravenesfot(B1) revenesfot(B3)
ravenesfote(C28)
Lit. B III, 143

Camedreos [camendreos, camedris, camandreos]
Teucrium chamaedrys L. 'Wall Germander'
?*Rhinanthus crista-galli* L. ? Yellow-Rattle
?*Pedicularis palustris* L. ? Red-Rattle
?*Veronica chamaedrys* L. 'Germander Speedwell'
gallice germandre(A1) germandre(A3) carewy*(A4)
erthclavere(B1) hertclavere(B3)
claveren(C2) anglice germander(C3) anglice med-ratil(C5) anglice mede- ratyl(C6) anglice claveren(C12) anglice clivere*(C18) germaundre(C21) germandrye(C27) med-retyll(C28) germandir lesse(C29) þe more orpyn*(C30) heor[t]clavre(C37)
Lit. B III, 133; Alph., p. 28, 1–9 & 152, 10

Camelea
Dipsacus sylvestris Huds. Wild Teasel
wlves-cambe(C37)
Lit. B III, 268f.

Cameleon [camelion, canelion, c. alba]
Dipsacus sylvestris Huds. Wild Teasel
(c. alba) wolvestasele(B1) wilde tasele(B3)
anglice wyld taisyl(C12) (c. alba) wulfthistill or wyld thistill(C21) (c. alba) wolffestasell(C28) hert-claver*(C28) wulvestasel(C37)
Lit. B III, 269

Cameleonta [camaleonta, cameliunca, camalconca, camaleoncia]

Cameleonta [camaleonta, cameliunca, camalconca, camaleoncia]
Dipsacus sylvestris Huds. Wild Teasel
Carlina vulgaris L. Carline Thistle
Cirsium acaulon (L.) Scop. 'Stemless Thistle'
Cnicus benedictus L. Blessed Thistle
anglice rennynwort(B2) anglice reningwort(B7)
rennigwort(C5) anglice remygwort .i. mente(C6) renneyngwort(C8) a þistel .i. rennyngwort(C10) a thistell .i. romyngwort(C11) gallice [sic] renewort, anglice crispmalowe*(C16) anglice renewort(C18) a thistill(C21, C24) mugwrt*(C23) a thystel .i. remyng-worton(C27) titemawe*(C28) rede mugwort*(28) renmingwort(C36) (alba) fowthistyll or qwhite-thistyll or seynte Mary thistell(C41)
Lit. Alph., p. 27, 30 ('Cameleonta alba, cardus coagulatus . . . herba est cum qua solet [coagulari] lac . . . angl. southistel') and note.

Camelia
Daphne laureola L. Spurge Laurel
lauriol(C28)

Camelopodion [camolopodion, camelipedion]
Marrubium vulgare L. White Horehound
anglice longwort*(B1)
horehound, maroil, houndisbery*(C21) lungwort*(C28)
Lit. André I, p. 67 (camelopodium) & II, p. 46

Camemelon [camelon]
Chamaemelum nobile (L.) All. Chamomile
Anthemis cotula L. Stinking Chamomile, Mayweed
Anthemis arvensis L. Corn Chamomile
maithe(B3)
mathyes(C2)

Camepetis [camapetis]
Ajuga chamaepitys (L.) Schreb. Ground-Pine
Cannabis sativa L. Hemp
henep(B3)
hempe(C12) henep(C37)
Lit. B II, 58 & III, 124

Camepitheos [camapiteos, camphiteos]
Ajuga chamaepitys (L.) Schreb. Ground-Pine
Rhinanthus crista-galli L. Yellow-Rattle
Pedicularis palustris L. Red-Rattle

anglice rachille(A1) rathhele(A3)
anglice osatille(B2)
medratele(C2) medaratell(C10) woderatell(C11) hen-penyse(C14) gallice et anglice osatille, anglice mederatele(C18) penygresse, medowratill(C21) rathehele(C22) anglice mederatel or ratyl or fartyng-grasse(C26) mederatel(C27)

Camesca
?
hyerwort(C28)

Camoleon [camelee]
Potentilla reptans L. Cinquefoil
synkfoile(C21) synkfoyl(C27)
Lit. André I, p. 85 (chamaezelos); N162

Camomilla [camamilla, camimella, camemilla]
Chamaemelum nobile (L.) All. Chamomile
gallice camamille(A1) camamille(A3)
camamille(B1) gallice et anglice camamille(B2) camemylle(B3)

Campana
?*Campanula* L. ssp. esp. *latifolia* L./*patula* L. ? 'Bellflower'
(c. cuculi, celi, agrestis) cokkowbelle(C1) (silvatica maior) þe grete wodebelle, (silvatica minor) pety wodebelle, wild sauge(C9) (c. cuculi, celi, agrestis) cokkowbell(C38)
Lit. see **Campania**

Campania
?*Teucrium scorodonia* L. ? Sage-Leaved Germander, Wood Sage
ambrose(C21)

Camforata [campherata, herba c.]
Artemisia abrotanum L. Southernwood
averoine, souþernewode(A3)
southernewode(B3)
gallice averoyne, anglice sowthernewode(C8) sothernwode(C21) avereyn, sotherwode(C24)

Camphora
laureaceous plant and/or its juice esp. *Camphora laurus* L./

Dryobalanops camphora L.
gallice canfre(A4)
gallice et anglice camphre(B2)
camfre(C2) anglice caunfre(C6) comfire(C28)

Cana
?*Capsella bursa-pastoris* (L.) Medic. ? Shepherd's Purse
anglice staunche(C19)

Cana Orti [c. ortensis]
Cannabis sativa L. Hemp
henep(C9)

Canabaria
Eupatorium cannabinum L. Hemp Agrimony, Holy Rope
wyld hemp or holy rope(C21) wylde hempe, holy rope oþer
don-netyll(C42)

Canicula
?*Filipendula ulmaria* (L.) Maxim. ? Meadow-Sweet
?*Potentilla anserina* L. ? Silverweed
?*Aphanes arvensis* L. ? 'Argentill', Parsley Piert
argentill(C21)
Lit. B III, 4f; Agnus Castus p. 212

Canicularis [caniculata, caniculare, herba c.]
Hyoscyamus niger L. Henbane
hennebane(A3)
gallice chenilé, anglice henbane(B1) gallice chenilé, anglice
hennebane(B1) hennebane(B3) henbane(B7)
anglice henbane(C6) henbelle vel henbane(C7) chenil, hennebane(C10)
henbane(C21)
Lit. N162

Canna
Arundo L./*Calamus* L. ssp. Reed
rede(C14)

Cannabis [canabs, canaps, canabus, canabum, canabe, canebus]
Cannabis sativa L. Hemp
Eupatorium cannabinum L. Hemp Agrimony, Holy Rope
gallice caunnvre, anglice henyp(A1) henep(A3)

gallice caumbre, anglice henep(B2) henp(B3, B5) hempe(B6) gallice camber(B9)
hempe(C1) (silvaticus maior) wilde hemp, holy rope (agrestis minor) lasse wild hempe(C1) (aquatica) anglice waterhenepe(C23) (agrestis, silvaticus) holy hemp(C38) gallice canfe, anglice hemp(C39)

Cannos
?*Chamaemelum nobile* (L.) All. ? Chamomile
camamylle(B3)

Capillus Veneris [capilli V.]
Adiantum capillus-veneris L. Maidenhair-Fern
anglice maydinher(A1) maydynheer(A3) anglice maydynher(A4) anglice maydenesher(B1) anglice maidenher(B2) maideneher(B3) (maior, terrestris) wyt crowratyl*(minor) maydynhere(C1) maidhere(C21) anglice maydenhere or waterwortte(C41)

Capparis [caperis]
Lonicera caprifolium L./*periclymenum* L. (Perfoliate)Honeysuckle
?*Convolvulus arvensis* L. ? Bindweed
anglice wodewynd(B1)
wethewynde(C6) wodebynde(C28)
Lit. B III, 264f.

Caprificus
Lonicera caprifolium L./*periclymenum* L. (Perfoliate) Honeysuckle, Woodbine
cheverefoyl, wodebynde, wethebyndde(C13)

Caprifolium [caprefolium, caperfolium]
Lonicera caprifolium L./*periclymenum* L. (Perfoliate) Honeysuckle, Woodbine
gallice chevrefoyle, anglice wetherwynde(A1) anglice wodebonde(A2) weyewynde(A3) anglice wodewynde(A6) anglice wldebinde(A8) anglice wodebinde(A9) anglice wodebynd(A12)
anglice hunisucheles(B1) gallice cheverefoil, anglice wethewynde vel wodebynde(B2) anglice wdebinde(B5)
anglice wodebynde(C1) cheverfoile, anglice wodebynde(C2) gallice chevebris, cheverfoyle, anglice wodbynde vel weþewynde(C6) gallice chevrebris vel chevrefoyl, anglice wodebynde(C8) anglice wodebyn(C16) wethywynde(C22) honysikill, wodebynde(C28)

Capsula
Capsella bursa-pastoris (L.) Medic. Shepherd's Purse, Sanguinary
estanche(B1)

Caput Canis [canis caput]
Antirrhinum orontium L. Small Snapdragon, Calf's Snout
houndeshed(B3)
cenocephalus, houndheed(C21) houndshede(C28) hundyshed(C32)
Lit. B 1, 88

Caput Leoni [c. leonachi]
Taraxacum officinale Weber Dandelion
dent de lion, liones toht(A3)
dent de leon, lyones toth(C22)

Caput Monachi
Taraxacum officinale Weber Dandelion, Priest's Crown
anglice prestes-croune(A5)
gallice dent de lion(B2)
dens de lioun(C6) gallice dent de lyon, anglice carteressage(C8) dent de lyon, lyontoþe(C10) den lyon(C21) lyuntoþ(C23) anglice dawndlyon or lyonstothe(C26) dent lyon, leonystoth(C27)

Caput Purgium
Delphinium staphisagria L. Stavesacre
staphisagre(C21) anglice lowsewort(C26)

Carabia
?*Carum carvi* L. ? Caraway
carawey(C28)

Carcocolla
Agrimonia eupatoria L. Agrimony
cordial, egrimonie(C21)

Cardamomum [cardemomum]
Sonchus L. ssp. Sowthistle
?*Elettaria cardomomum* (L.) Maton ? Cardamom
anglice fouthystyll(C3) anglice erthnute*(C23) anglice tame cardomonye(C26)
Lit. N163

Cardiaca
?*Leonurus cardiaca* L. — ? Motherwort
?*Alliaria officinalis* Bieb. — ? Hedge Garlic
cardiacle, oylfere [corr. cilspere](C41) cardiak(C21) cardiacle(C38) anglice cardyake or cilspere or peronell(C41) cardiake, cilsper(C42)
Lit. Agnus Castus p. 218

Cardo
Dipsacus sylvestris Huds. — Wild Teasel
anglice wilde tasel, gallice cardoun savage(B2) anglice þestel(B2) potþistel(B5)
(c. minor, rubea, maculata) wolvesthistell(C1)
(c. fullonis) tesyl, wolvescombe(C1) tesylle(C6) (silvestris) wylde tesyll(C7) gallice gardoun, anglice wild tasill(C16)

Cardo Asininus [c. asinus]
a thistle, poss. *Onopordon acanthium* L. ('Carduus nutans')
þistel(A3)
lasse asse-thistil(C1) a þystel(C10) thystyle(C22) scariol(C36)

Cardo Benedictus [c. benedicta, cardus b.]
Cnicus benedictus L. — Blessed Thistle
Senecio vulgaris L. — Groundsel
soweþistel(A3) groundesuelie(A7)
gallice cardoun beneit, anglice foutþistel(B2) groundeswelie(B3) anglice fauthistel(C5) voweþistell(C10) verchardon, fouthistill(C16) anglice foughthystel(C18) voweþistil(C19) groundswille(C21) ffowethystyle(C22) anglice wode-tesyl [corr. -thystel] or Our Lady thystel(C26) wowþistill(C31) grundeswele(C32)

Cardo Maculatus
Sonchus L. ssp. — Sowthistle
thowthistill(C21)

Cardo Media
Sonchus L. ssp. — Sowthistle
þistel(C35)

Cardomum [cardami, cardomi, cardomon, cardami, cardamon, cardamonum]
Rorippa nasturtium-aquaticum (L.) Hayek — Watercress
Lepidium sativum L. — Garden Cress

tuncarse-seed(C10) touncrasses(C11, C24) touncresses(C21)
touncressez(C27) watercresse(C28) cres-seede(C39)
Lit. N163

Cardonum
?*Dipsacus fullonum* L. ? Wild Teasel
wild tasill(C21)

Cardus Maior
? a thistle
grant cardon, anglice wodcle[te?](B1)
tasell(C28)

Cardus Minor
?*Cnicus ferox* L. ? 'Cruel Thistle'
petit cardon, smal thistile(B1)
smale thistell(C28)
Lit. B III, 210

Cardus Sancte Marie
Silybum marianum (L.) Gaertn. Milk-Thistle
anglice fowthistel or Saint Mary mylke-thistel(C26)

Cardus Silvaticus
? a thistle
wilde þistel(A11)
Lit. B III, 268

Carduus [cardus, cardo]
Sonchus oleraceus L. Sowthistle
Dipsacus fullonum L. Teasel
anglice fauthystel(A5)
tassile(B4) tesul(B5)
(c. communis albus) tesile (c. minor albus) vollerystesyl (c. magnus)
asse-þystyl þe more(C1) (albus) anglice tesyll(C26) brod-þistel(C39)

Carduus Fullonis [cardus, cardo f.]
Dipsacus fullonum L. Teasel
Saponaria officinalis L. Soapwort
crowesope(C2) anglice tasyll(C3) crowsope(C7) savvenere,
crousope(C10) cavvener, crouesop(C24) savvoner, crowsope(C27)

Carectum
Trapa natans L. Water Chestnut
castanea marina, anglice waternot(C26)

Carei Porcini [corr. pontici]
Daucus carota L. Wild Carrot
dauek(C10) dauke(C11) dauk(C21, C27)

Careon [carionus, cariona]
Corylus avellana L. Hazel, Cob-Nut
haselnote(C9)

Carex
Carex L. ssp. Sedge
flag, ryssches, scharp ryssches(C1) segges(C29) anglice snydel(C36)

Cariatarum [for Caricarum]
Ficus carica L. Fig
figes seches(A5)

Carica
Ficus carica L. Fig
anglice wodeberie*(B1)
drye fyges(C2) drye fyge(C9) drie figges(C21) pasnepe*(C21)

Caride Miristice
Myristica fragrans Houtt. Nutmeg
anglice notemuge(C10) .i. nutmygge(C11) notmuges(C24)

Cariota
Daucus carota L. Wild Carrot
Pastinaca sativa L. Wild Parsnip
[f]ellemore(B1)
Lit. B I, 59

Carita
?
hemlok, wodeþistill [corr. wodewistill?] (C28)

Carma
Anthemis cotula L. Stinking Chamomile, Mayweed
gallice amereite, mathe .i. maidenwed(B1)
maithe(C28)

Caro Leonis
Cinnamomum zeylanicum Blume — Cinnamon
Cinnamomum cassia Blume — Cassia, Bastard Cinnamon
anglice canel(B7)
anglice canel(C5, C6, C8) anglice canelle(C13)

Carocaria [for colocasia]
Mentha L. ssp. esp. *Mentha longifolia* (L.) Huds. — Horse-Mint
haresment [sic](B1)
Lit. B III, 123f

Carpobalsamo [carpo bassamo]
Teucrium chamaedrys L. — 'Wall Germander'
gallice germaundre(B2)
anglice medratil, gallice germandre(C16) anglice moderatele, gallice germandre(C18)

Cartadona [cartadana]
?*Citrullus colocynthis* Schrader — Colocynth
coloquintida, wild nepe(C21)

Cartamus
?*Carthamus tinctorius* L. — ? Safflower, Bastard Saffron
?*Crocus sativus* L. — ? Saffron
crocus almanicus, safron of Almayne(C1)

Carteron
'Creta marina'?
centinodie(C1)
Lit. see **Cretanus Marinus**

Caruta
Pastinaca sativa L. — Wild Parsnip
?*Sium latifolium* L. — ? Water Parsnip
?*Berula erecta* (Huds.) Coville — ? 'Narrow-Leaved Water-Parsnip'
wilde skyrwhit(C13)

Carvi
Carum carvi L. — Caraway
(c. domesticus, maius, ortolanus) carwy(C1) sic vocatur ȝarow*(C14)

Cassamus
Conopodium majus (Gouan) Loret	Pignut, Earthnut
Cyclamen hederifolium Ait.	Sowbread
anglice erth-apples or erth-nottys or wyld dyll(C26)
Lit. N165

Cassia
[*Cinnamomum cassia* Blume]
gallice lavendre*(B1)
lavendre*(C28)
Lit. Alph. App. p. 208 ('.i. hederae semen')

Cassia Fistula [cassea f.]
Cinnamomum cassia Blume	Cassia, Bastard Cinnamon
casseafistle(A3)
anglice cassiafistre(B2)
þe fruyt of a tree, bot cassia lignea is þe rynde of þe frute(C10) gallice cassialigne, anglice cassiafistre(C16)

Cassia Lignea [cassea l.]
Cinnamomum cassia Blume	Cassia, Bastard Cinnamon
cassealigne(A3)
gallice cassiligne(B7)
gallice cassealigne(C6) gallice cassialigne(C8)

Cassilago [cassialago]
Hyoscyamus niger L.	Henbane
gallice geliner, anglice henebel vel heneban(A1) hennebane(A3) anglice hennebane, gallice geliner(A4) hennebane(A7)
gallice genele vel geluner, anglice hennebelle(B2) hennebane(B3) henebane(B5)
anglice henebane, gallice gemel vel gelmoth(C5) anglice hennebonewort(C20) anglice heqwale[?](C23)

Castanea
Castanea vesca Gaertn.	Sweet Chestnut
castaine(A3)
gallice et anglice chasteynes(B2)
chastayne(C2) anglice casteyne(C6)

Catapucia [catepusia, cataputium]
Euphorbia lathyrus L.	Caper Spurge (seed of)
gallice cacapuce .i. la semence de spurge(A1) catapusce, spurge-sed(A3)

anglice spurge(A8)
gallice catapus, semence de spurge, anglice spurgesed(B2) spurge(B4) anglice spurge(B5)
spourge(C3) gallice catapus, anglice spurgeseed(C6) gallice catepuse, anglice spurgesed(C8) anglice faythorwort(C32) romani vocant eam spuroge, anglice sperwort*(C43)

Catarian
?*Coronopus squamatus* (Forsk.) Aschers. — ? Swine-Cress
anglice hertishorn(C12)

Catos
?*Rosmarinus officinalis* L. — ? Rosemary
rosemarin(C28)

Cauda Caballina
Equisetum L. ssp. — Horsetail
(c.c. minor, c. equi minor, c. pulli) horstayle, coltystale(C1)

Cauda Equina [c. equi]
?*Digitalis purpurea* L. — ? Foxglove
?*Equisetum* L. ssp. — ? Horsetail
gansole, voxesglove(A3)
gallice gauntele, anglice foxestonge(B1) gallice ganxele, anglice foxesglove(B2) gallice gaunsil, anglice foxglof(B7)
foxegloven(C2) gallice gaunciel, foxeglove(C6) gauncel, anglice foxeglove(C8) foxesglove(C10) anglice foxegloven(C12) gouncel(C13) gallice ganxele, anglice foxglove(C16) glansole, anglice voxésglove(C22) anglice horstayle or coltistayle(C26) foxesglove(C27) anglice foxglove et secundum alios schavegras(C36) is like trefoyle(C37) similis trifolio, anglice horsegres(C39)

Cauda Lupina
?*Lithospermum officinale* L. — ? Gromwell
Verbascum thapsus L. — Mullein
gromyl(B3)
wolfestayle(C1) anglice gromyle(C12) anglice molyn or flotwort [corr. feltwort](C26)

Cauda Porcina
Lithospermum officinale L. — Gromwell
Peucedanum officinale L. — Hog's Fennel

Sonchus oleraceus L. Sowthistle
gallice grommyle(A1) gromoil(A3)
gallice et anglice gromil(B2)
anglice gromyle(C6) greynmyle, cankirworth(C9) gromile(C13) cardoun, anglice scharpe thistill(C16) gromel(C21) wormsede, swynesthistill(C21) anglice dog-fenel or mathegyn or wormesede or matheyn(C26)

Cauda Pulli [c. pulpi]
Equisetum L. ssp. Horsetail
padokpipe(A2) anglice paddepype(A6) anglice padepipe(A8, A9, A12)
wodepipe(B3) anglice padpipe(B9)
padokepype(C3) paddokespipe(C10) anglice wodpipe(C12) padpipe(C15) cauda caballina, folistayle, paddokpipe(C21) a padokpipe(C24) poddokpype(C27) folistayle(C32) paddpype(C34) paddepipe(C37) foletayle(C41)

Cauliculus Agrestis
?*Brassica oleracea* L. ? Wild Cabbage
?*Peucedanum ostruthium* (L.) Koch ? Master-Wort
?*Silene vulgaris* (Moench) Garcke ? Bladder Campion
(rubeus) rede crowratyl [corr. cowratyl] [albus] wyt crowratyl [corr. cowratyl](C1) cou-ratyll(C21) meysterwort(C26) (rubeus) red crowratyll [corr. cowratyll](C38) anglice glaunde or kowrakke [corr. kowrattle](C41)
Lit. Agnus Castus p. 219 (also with the reading cowrake)

Caulis [caulus]
Brassica oleracea L. Cabbage
Sinapis arvensis L. Charlock, 'Wild Mustard'
anglice coul(B2)
(c. alba, ortulana) wyld caulwort (c. rubea) red caulwort (c. alba/rubea domestica gallica) frenche wort (c. romana) wort (c. gallica rotunda) caboche (c. agrestis alba) wyld wort(C1) cowlwortes(C2) anglice colewort(C12) anglice coule(C16) (c. agrestis) carlok (c. gallica) caboche (c. domestica) wort(C21) (rubia c.) rede cole(C28) (gallica) caboche, french worte(C32) (alba) whytt wertys (rubia) red wort (alba domestica gallica) whitt frensch wort (rubia domestica gallica) red frensche wort(C38) (gallica) anglice kaboche or frensche wortte (agrestis) wylde wort or wylde kole(C41)

Celidonia
Chelidonium majus L. Greater Celandine
Glaucium corniculatum (L.) Rudolph 'Red Horned-Poppy'

Glaucium flavum Crantz 'Yellow Horned-Poppy'
gallice celidoyne(A1) celidonie(A3) gallice celidon(A8)
celidoyne(B1) gallice et anglice celidoyne(B2) celidoyne(B3)
selidoyn(B6)
celidonye, tetyrwort (minor, agrestis) wylde celydonye(C1) (agrestis) celidonye de meer(C2) anglice celidone(C3) moun celidoyn de mere(C10) (agrestis) menta [corr. memita], gallice celydoyne de mer, anglice mente [corr. memite](C12) anglice skallewort(C20) celidony or tetyrwort(C21) (agrestis) memithe, wild celidonie(C21) anglice heihovi*(C23) celidony de mer(C27) celidony or celandine . . . wylde celendyne yclepid memithe(C31) romani vocant chilizene, anglici celidoyne(C43)
Lit. Alph., p. 36, 5ff

Celsa [celsis, celsus]
Morus nigra L. Mulberry
mulberitre(A5)
malberytre(C3) spica celtica, mollebery(C10) mulbery(C27) þe most morell(C28)

Celsus Amara
?*Blackstonia perfoliata* (L.) Huds. ? Yellow-Wort
?*Centaurium erythraea* Rafn ? Common Centaury
major centorie(C24)

Celtica
var. spiked plants esp. *Valeriana celtica* L.
anglice brom(B1)
(c. spica) anglice calketrappe(C6)

Ceme
Rumex aquaticus L. Red Dock
Rumex sanguineus L. 'Red-Veined Dock'
reed dokke(C21)

Cenophalion [for cenocephalion]
?*Teucrium chamaedrys* L. ? 'Wall Germander'
anglice hertclavere(B1)
Lit. B III, 133

Centaurea [c. maior, c. minor]
Centaurium erythraea Rafn Common Centaury
Blackstonia perfoliata (L.) Huds. Yellow-Wort

gallice centoré, anglice cristesladdir(A1) centoyre, cristesladdre(A3) urþegalle(A3) gallice centor[ie](A5) anglice seynte Marie laddre(A6)
(minor) anglice erthegalle (maior) anglice herdwort(B1) (maior) gallice centorie, anglice cristesladdre (minor) anglice horssegalle(B2) centorie(B6)
(maior) þe more centorye (minor) erþegalle(C1) ertheyalle or cristesladder(C2) anglice centory(C3) (maior) gallice centorie, anglice erthegalle vel cristesladdren (minor) febrifuga, anglice feþerfoye vel horsgal(C6) (minor) gallice fetherfoy, anglice horsgalle(C16) (minor) gallice fevervoye, anglice horsgalle(C18) (maior) more centory (minor) cristsleddre, less centory(C21) (minor) fetherfoye(C23) crystesledder(C27) (maior) herdwert (minor) erdgall(C28) feþerfoye(C31) (maior) heortgalle (minor) fetherfoie(C37) herrehialle(C43)
Lit. B III, 84 & 134

Centinaria [centenaria, centinervia]
Plantago major L. — Plantain, Waybread
Plantago media L. — 'Hoary Plantain'
lambestonge(C10) centoys vel centory(C11) lambistunge(C27)
Lit. Agnus Castus p. 234

Centinodium [centinodia, centenodium]
Polygonum aviculare agg. — Knotgrass
Vicia L. ssp. — Vetch
gallice ternue*, anglice quych(A6)
swyncrasse(B4) contnod(B6)
sentinodie, sparwt[ong](C1) anglice quichon vel swynsgrasse vel onefortrede(C5) anglice quichene vel swynysgres vel onfortrede(C6) anglice swynesgresse vel cnottydgresse(C12) quicchene, swynisgras, (c)onfortrede(C13) gosgrasse*(C14) sparwistung, swynesgrece(C21) stichegresse(C21) anglice quych or knotgras or sparawtong or hoggsgrasse(C26) swyngras vel onfertrede(C36)
Lit. B III, 241

Centonica [centonia]
Artemisia maritima L. — 'Sea Wormwood'
centonigla, merlynge [corr. amer aloynge?], flewort*(B4)
anglice seewourmod(B2)
anglice seewormoth(C6) wermode(C10) anglice seewormoud(C18) wyrmode(C21) sewurmod(C23)

Centonicula
?*Pulicaria dysenterica* (L.) Bernh. ? Fleabane
?*Pulicaria vulgaris* Gaertn. ? 'Small Fleabane'
?*Plantago indica* L. ? Fleawort
Cristoforus vel policaria, romani vocant eam meraloigne [corr. amer aloigne?], anglici flewort(C43)

Centrum Galli [c. galla]
Salvia pratensis L./*verbenaca* L. Clary
Agrostemma githago L. Cockle
Lolium temulentum L. Darnel
centregalle(A3)
gallice centrogalle, anglice scallewort(B2) clareie(B3)
anglice scalleworth(C6) nele, cokell, popell(C7) centirgalle(C10) gallice neele, anglice cockel, alibi centrum galli . . . henebane*(C12) scallewyrt(C13) gallice centrogal, anglice schalwort(C16) gallice centrogalle, anglice scallewort(C18) wyld oculus Christi(C26) schallewort(C36)
Lit. Alph., p. 38, 15

Centrum Pulli [? error]
? ?
gallitricum, lambistong(C24)

Centum [for centinervia?]
Plantago major L. Plantain, Waybread
planteyn, weybrede(C21)

Centum Capita
Allium ursinum L./*vineale* L. Ramsons/Crow Garlic
Allium schoenoprasum L. Chives
anglice maudefeloun*(A6)
ramese, crowgarlyk(C6) affodyll(C7) anglice wilde garlek(C8) affodille, civys(C9) affadille(C21)
Lit. N167

Centum Grana
?
camemylle(C21)
Lit. cf. Alph., p. 38, 15

Cepa [MS cera]
Allium cepa L. Onion
hennelek(B1)

Cepe [cepa, sepe]
Allium cepa L. Onion
gallice oynnun(A1) oynun(A3) oynnun(A3)
anglice oynyonnes(B1) gallice et anglice oynouns(B2) oynouns(B3)
anglice oynoun(B7)
oynoun(C1) (c. canina, agrestis) wylde oynon (c. silvestris) wylde oynnon(C1) gallice [sic] hewlyke [corr. hennlyke], anglice unouns(C16)

Cepe Marina [cepa m.]
Scilla maritima Baker Squill
squill(B4)
squilles(C21) swilles(C40)

Cepula
Allium schoenoprasum L. Chives
syve(C21) chibol(C29)

Cerasa [ceresa, cecera]
Prunus cerasus L. Cherry
cericer, chyrise(A3)
gallice cerice, anglice chirs(B2)
chersen(C2) anglice chyrien(C6) anglice hertestonge*(C8) anglice chersen(C12) gallice ceruse [sic], anglice chirise(C18) cheri-beri(C23)

Cerfolium
Anthriscus cerefolium (L.) Hoffm. Chervil
gallice cerfoylle, anglice villis(A1) cerfuyle, villis(A3) gallice cerfoyle, anglice villis(A4)
cheverefoile*, cheverille(B1) gallice cerfoil, anglice chervel(B2)
hertestonge*(B5) cerfoyle, ville(B8)
cerfoyle(C1) anglice cerfoyle vel villes(C6) cerfoile, crowfote*(C21)

Cervina [c. lingua, lingua c.]
Phyllitis scolopendrium (L.) Newm. Hart's-Tongue Fern
gallice cerflang, anglice hertestong(B7)
gallice cerflange, anglice hondystong*(C6) hertistonge(C13)

Cervipodion
?*Conopodium majus* (Gouan) Loret ? Pignut, Earthnut

hertisfote(C21)
Lit. Alph., p. 40, 26

Cestonia
Artemisia absinthium L. Wormwood
anglice wormoth(B7)

Cestres
?*Acanthus mollis* L. ? Bear's Breech
brank-ursine(C21)

Cestrios
?*Chelidonium majus* L. ? Greater Celandine
tetrewort(C28)

Cestros [cestros maior]
Sempervivum tectorum L. Houseleek
housleeke(C21)

Cestrum
Betonica officinalis L. Betony
betoyne(C10, C27) betonye(C11)

Chatambia
Brassica L. spp. Cabbage
c'est chous blanc(C17)

Christiana [herba c.]
Lathyrus L. ssp. Vetchling
?*Senecio vulgaris* L. ? Groundsel
cristyane, an herbe þat haþe a stalke out of þe whyche go þe oþir stalkys with rodende levys as posis and yclepyd ursope(C31) anglice grondswyly(C36) crystayne(C38) similis est amarusce .i. marche(C39)
Lit. Alph., p.194

Christicius
Althaea officinalis L. Marsh Mallow
holihok, wild malwe(C21)

Cibus Anatis
Stellaria media (L.) Vill. Chickweed
Lemna L. ssp. esp. *L. minor* L. Duckweed

chekemete(C2) anglice schekemete(C12) enmete(C21)
Lit. Alph., p. 97, 1

Cibus Porcinus
Cyclamen hederifolium Ait. Sowbread
Conopodium majus (Gouan) Loret Pignut, Earthnut
erþenote(C12)
Lit. B III, 84f.

Cicer [cizera]
Cicer arietinum L. Chick-Pea
Pisum sativum L. Pea
anglice ciches(B7)
(c. sanguineus, niger) chyche (c. albus) chyche þe whyte(C1) anglice chiches(C5) chiches(C6) french pese(C21) frensche pese(C21)

Cicer Tuscicum [corr. rusticum?]
?*Thymus serpyllum* L. ? Wild Thyme
peletre(C21)

Cicerum [ciceri]
Lathyrus L. ssp. Vetch
.i. fiches(A5)

Cicida
Bryonia dioica Jacq. Bryony
?*Ecballium elaterium* A. Rich. ? 'Squirting Cucumber'
wild nepe(C21)
Lit. Alph., p.170 (sicida)

Ciclamen [ciclaminus]
Cyclamen hederifolium Ait. Sowbread
Conopodium majus (Gouan) Loret Pignut, Earthnut
anglice eorþenote(A1) urthnotte(A3) erþenotte(A3) anglice extat tos coluoranis(A6) aþnote anglice(A8)
anglice erthnote(B1) erthenotes(B4) anglice erthnote(B5) anglice yornote(B9)
erthenotes(C2) anglice ȝornotte(C3) noys de terre, anglice erþenote(C10) erdnote, dylnote, halsewort*(C21) noys de tere, erthenote(C27) herdnote(C28) swynesnep(C28) piggisbrede, erþenote or hertnot(C31) anglice herþontes [sic](C32) hoggesmete(C39) anglice erthnotte or dilnotte or slite or haylewortte(C41)
Lit. B III, 84f; Agnus Castus p.142, 24ff

Ciclania
?*Cyclamen hederifolium* Ait. ? Sowbread
anglice swynesnepe(B1)
Lit. *Anglia* 95(1977)116 ('rapum porcinum')

Cicorea [cicora]
Cichorium intybus L. Chicory
conf. with *Calendula officinalis* L. Marigold
gallice cicorie, anglice roddis(A1) cicorie(A3) gallice cicorie(A4)
gallice cicorie, anglice hardhewe(B2) cicorie(B5)
anglice marigolde(C3) gallice sicorye, anglice marygold vel redfloure vel houene(C6) heuene(C5) gallice cicory, anglice hardhew vel solsecle(C16) gallice cicorie, anglice hardhewe(C18) anglice gooldeword(C20)

Cicuta
Conium maculatum L. Hemlock, Herb Bennet
gallice herbe benet, anglice hemelok(A1) herbe beneite, hemeloc(A3) herbe benet(A4) anglice hemeloc(A6)
gallice herbe benet, anglice humlok(B1) woddewort, pimpernel(B1) gallice herbe beneyt, anglice humeloc(B2) gallice herbe benet, anglice hemeloc(B5) herbe benett, humblock(B6) gallice herbenet, anglice hombelonk(B9)
hemlok(C1) clofþounke*(C2) anglice homlok(C3) cloftunke*(C7) humlok(C21) (nigra) anglice hornwhistyll(C41)
Lit. Alph., p. 40, 3

Cimbalaria [cinbalaria]
Cymbalaria muralis Gaertn., Mey. & Scherb. Ivy-Leaved Toadflax, Pennywort
Umbilicus rupestris (Salisb.) Dandy Pennywort
anglice peniwort(A2) anglice penigres(A2) deneriol(A3) denarole, anglice peniwort(A4) anglice peniwrt(A6) penigres(A8, A12) anglice penigresse(A9)
gallice denariole, anglice penywith(B1) gallice denerole, anglice peniwort(B2) anglice penigres(B5)
anglice penygrysse(C3) gallice daneliol [sic](C16) deveriol [sic](C22)

Cimbaria
Melilotus officinalis (L.) Pall. Melilot, Sweet Clover
Trifolium pratense L. Red Clover
anglice honisoukel(B9)

Cimbrium
?
mugwort(C21)

Cime Ebuli
Sambucus ebulus L. Dwarf Elder tops
gallice coperun de eble, anglice crop of walwert(C20)

Cime Rubi
Rubus fruticosus L. Bramble tops
ciouns de runce(C7) ciuons de rons, bremelcroppe(C10) gallice cyowns de rownce, anglice brembellcroppys(C12) brymmelcrop(C21) þe crop of þe brembyll(C24) cyuons de rounse, brimelcrop(C27)

Cime Rumicis
Rubus L., or *Rosa canina* L. Briar tops/tops of the Dog Rose
gallice coperuns de rounce, anglice crop of þe red brere(C20)

Ciminum Barbarum
Siler montanum Crantz 'Sermountain'
Laserpitium siler L. 'Sermountain'
anglice sylermontayn(C26)

Cinnamomum [cinomomum]
Cinnamomum zeylanicum Blume Cinnamon
Cinnamomum cassia (L.) Blume Bastard Cinnamon, Cassia
canel(A3) gallice canel(A4)
gallice et anglice canel(B1) canele(B3) canell(B4) canel(B5)
canele(C2)
Lit. N168

Cinoglossa [cinglossa]
Cynoglossum officinale L. Hound's Tongue
gallice chenelonge, anglice hundistunge(A1) chenlange, hundestunge(A3) anglice hundetung(A5)
anglice hundistong(B1) gallice chenlange, anglice houndestonge(B2) cerlaunge*(B5)
anglice huntetong(C3) anglice hondestong-lef(C6) hundistonge, [?]algail(C13) houndistonge(C28) hondstong-lef(C36)

Cinomia
?Plantago indica L. ? Flea Seed, Psyllium (Plantain)
prosill(C21)

Cinus
Crataegus monogyana Jacq. Hawthorn
Pistacia lentiscus L. Mastic
hauþorne, hauþorne-tre(C6) lentiscus, lynde(C9)

Ciperum [cipperium]
Centaurium Hill ssp. Centaury
treis maner de centorie(B2, C18)
foille, erthegalle(C22)

Ciperus
Cyperus longus L. Galingale
?*Acorus calamus* L. ? Sweet Flag
galangal or the grete seg(C21) the rote of englyssh galyngal(C26) þe eggid gras(C31)

Cirisus
?
borage(C21)

Ciroglossa
Phyllitis scolopendrium (L.) Newm. Hart's-Tongue Fern
hertistong(C21)

Cisamus [cisama]
?*Apium graveolens* L. ? Wild Celery, Water Parsley
waterpersil or wilde persil(C9)

Cisson [cisona, celson, celsona, cusor]
Hedera helix L. Ivy
yvy, edre(C9) c'este .i. herbe qui aporte la gomme que ad a non popenac(C17) yvy-tree(C28)

Cissor
Galium aparine L. Cleavers
anglice clivere(B1)

Cito Valens [cetivalens, cetevalens]
Curcuma zedoaria Rosc. Setwall, Zedoary
cedewal(A3)
gallice et anglice cetewale(B2)
gallice et anglice ceduall(C16) cedewale(C22) zedwale(C31)

Citonaletis
?
centenarie(C23)

Citonia [citonica, citonium]
Cydonia oblonga Mill. Quince
coynz(A3) coyns(A5) gallice coyns(A5)
coyne(B7)
quincis vel queynis(C31)

Citonius
Cydonia oblonga Mill. Quince Tree
quynce-tre(C9)

Citrago
Melissa officinalis L. Balm, Lemon Balm
gallice pigmentarie, anglice piment(A1) pigmentarie, piment(A3)
anglice piment vel modworth(B7)
anglice pygment vel medeworth(C6) pyment, medeworth, bourfayr(C13)
baume(C21) pigmentari, pyement(C22) anglice bame(C26)
Lit. Alph., p. 40, 29 ('bonrefair') & 115, 11

Citragulus [citragula, citranguli]
Citrus aurantium L. Orange
orenge-tre or orenge appul-tre(C9) poma de orynge(C36)
Lit. N169

Citrangulum
[*Citrus aurantium* L.]
febrifuga, feþerfoiȝe*(B3)

Citruli
Citrullus lanatus (Thunb.) Mansf. Water Melon
anglice cerlus(C5) gallice cerlus(C6, C8) cherles(C13) [all errors for citrules?] cerlus(C36)
Lit. N170

Citrullus
Citrullus lanatus (Thunb.) Mansf. Water Melon
citrull fetose(C1)
Lit. N170

Clamer [?]
?*Cyclamen hederifolium* Ait. ? Sowbread
panis porcinus, urthinote(C22)

Claretum
Salvia sclarea L./*horminoides* Pourr. Clary/Wild Clary
gallice et anglice clarry(C16)

Coccinellus
Prunus domestica L. Plum
plum-tre(C9)

Coconidium [cossonedium, coconitidum]
Daphne gnidium L./*laureola* L. Fruit of the Laurel
semence de lauriol(A3)
laureole-sed(B3) sede of lauriol(B6)
gallice baye de laureole(C20) ye seed of laureol(C21)

Codium
Papaver L. ssp. Poppy
blac piper*(C2) anglice chesbollsede(C3) black popy(C26)

Colicus
?*Carlina vulgaris* L. ? Carline Thistle
efodrote(C28)
Lit. B III, 81

Collocasia
Mentha L. esp. *Mentha longifolia* (L.) Huds. Horse-Mint
Nepeta cataria L. Catmint
anglice horsmynte(B5)
katmynte(C1)
Lit. cf. André I, p. 96 (colocasium) & II, p. 71; B III, 123f.

Colopendrya
Phyllitis scolopendrium (L.) Newm. Hart's-Tongue Fern
also appl. to *Adiantum* L. & *Asplenium* L. ssp.
saxifrage(C9)

Coloquintida
Citrullus colocynthis Schrader Colocynth
herswet(B1) anglice coloquintide(B2)

anglice wylde nep(C6) campana cuculi, cocowbelle*(C21) smerenep, it is a appill(C28)
Lit. André I, p. 97 & II, pp. 71f.

Colubrina
Polygonum bistorta L. — Bistort
Dracunculus vulgaris Schott — Dragon Arum
nedderwort(B4)
edderwort, dragaunce(C2) anglice dragance(C3) nedderwort or dragance(C21) anglice serpentyne or dragans or addirwort(C26)
Lit. André I, p. 97 & II, p. 72; N170

Columbaria
Dracunculus vulgaris Schott — Dragon Arum
serpentine, nedderwert(A3)
serpentarye(C12) anglice moderwort [corr. nederwort](C22)

Columbina [columbaria]
Aquilegia vulgaris L. — Columbine, Culverwort
?*Geranium columbinum* L./*molle* L. — ? Cranesbill
columbine(A3) gallice columbin(A8)
gallice columbine(B1) gallice columbine, anglike hauekesfet(B2) anglice columbine(B5)
columbine, cokkesfote, dovefote, dowvewort(C1) columbine(C2) gallice columbyne, anglice hauekesfot(C8) columbine, cokfoot, dowort(C21) kokysfote(C32) anglice columbyne or cokfote or dowewort(C41)

Coluragium
Polygonum hydropiper L. — Smartweed, Water-Pepper, Red Knees
anglice rede knos(B7)
anglice redelnes [corr. rede knes](C6) anglice rede dokkes*(C8)

Comiticiale [for comitiale; comisale]
Gentiana L. ssp. — Gentian
gencian, baldmony(C21)
Lit. André I, p. 98 (comitialis) & II, p. 73; Alph. App., p. 208

Condrisia
Origanum dictamnus L. — Dittany of Crete
Dictamnus albus L. — Bastard-/White Dittany
anglice dytteyne(C26)
Lit. André 1, p. 98 (condris(s)um)

Consolida
Bellis perennis L. Common Daisy
Symphytum officinale L. Comfrey
bonworte(B4) baynwort, comferie(B6)
comferie, wringinwort, daiziee(C10, C11)

Consolida Maior
Symphytum officinale L. Comfrey, Consound, Bruisewort
and conf. w. consolida media and minor
gallice confery, anglice muche brisewort(A1) la grant consoude, muche bryswort(A3) gallice confery(A4) anglice counferie, anglice briswrt(A6) anglice conferie(A8) gallice conferye(A9) anglice confurie(A10) gallice confirye(A12)
conseude, anglice homund vel speringwort(B1) gallice la grant consoude, anglice michele brosewort(B2) conferie(B3) brusworte(B4) anglice confirie(B5) comfirie(B5) hayreth*(B6)
confyry, consoude (c. minor rubea) anglice confyry(C1) confirie vel brusewort(C2) conferye or brusewort(C2) anglice comfery(C3) gallice comfery, anglice brysworth vel bo[n]worth(C6) bonwort(C8) anglice micheliswrt(C23) conferre, wrengingworth, daiesyeu(C24) springwort, confery(C28)

Consolida Media
Bellis perennis L. Daisy
Chrysanthemum leucanthemum L. Ox-Eye Daisy
Chrysanthemum segetum L. Corn Marigold
?*Ajuga reptans* L. ? Bugle, 'Helfringwort'
?*Symphytum tuberosum* L. ? 'Tuberous Comfrey'
anglice elferingeworth(A1) la meine consoude, elferingewort(A3)
gallice la mene consoude, anglice helfringwort(B2) medworte(B4) daysheie(B5) rode(B5)
(c.m. alba, centrix) whyt-gould, boþel (c.m. crocea) ȝelw boþel(C1) whit bothel(C2) medewort(C6) anglice boseleye[?](C8) wryngynwort, middel consoude(C10) anglice whyt bothel vel medwort(C12) medewrt(C13) quiȝt goolde(C21) anglice bonwort(C23) wryngyngwort, medyl consoude(C27) elinswort(?)(C38) daysee(C39) wryngenge-wort, medyl consounde(C40) wryngensworte(C40) qwhite bothon or qwhite golde(C41)
Lit. B I, 13f., II, 7; III, 15

Consolida Minor
Bellis perennis L. Daisy
and conf. with Consolida maior/media

deyseye(A1) la petite consoude, dayse(A3) anglice daysies(A4) anglice daesheye(A6) dayseue(A10)
la petite consoude, anglice daieseye(B2) daiseie(B3) la peti consoude, et est bonworte idem est dayseye(B4) daysheie(B5) daysy(B6) consaude(B8) gallice consout, anglice bainewort(B9)
anglice comfery, dayses(C3) anglice daysye(C6) daysey, uenwort, bonewort(C7) dayizie(C12) bonwrt, anglice daysye(C13) gallice le pite [corr. peti] consuude, anglice daysye idem quod bruswort(C16) anglice plaggell* idem peniwrt*(C23) brusewort(C27) daisey þat knetteþ broken bones(C37) anglice brotherwort*(C40)

Copiosa
?
herbe copie(C10, C40) herbe copye(C11) herb copi(C24) herbe copye(C27)

Cordimena [cardumene, cordimen, cordumen, cordumeni]
?*Carum carvi* L. — ? Caraway
?*Cuminum cyminum* L. — ? Cumin
budthistel*(B5)
anglice wild comyn or caraway(C26)
Lit. Alph., p. 30, 20; see **Bombix**

Coriandrum [coliandrum]
Coriandrum sativum L. — Coriander
gallice coriandre(A1) coriandre(A3)
gallice coriaundre(B1) gallice coriaundre, anglice coliaundre(B2) coliandre(B3) coliaundre(B5) coliandyr, daysie*(B8)
coriandre, coliandre(C1) coliandre(C2) gallice cherendir [sic], anglice coliandre(C16) anglice coryander or colyander(C26) coliandir, flewort(C31)

Corigialia
Convolvulus arvensis L. — Bindweed
bynde de urse, anglice berbynde(B1)
Lit. B III, 27

Coringo
?*Anchusa arvensis* (L.) Bieb. — Bugloss
?*Picris echioides* L. — 'Bristly Ox-Tongue'
lang de bof, oxtung(C21)

Coripe [for corpore?]
?
liverwrt(C23)
Lit. see **Corpore Epatica**

Cormeon
?
lyly(C21)

Cormindi
Hedera L. ssp. Ivy
ivi-tre(B1)

Corneria
Melissa officinalis L. Common Balm
medwort(B5)

Cornopedion [cornopodiom, cornipedium, cronapedion]
Ranunculus L. esp. *Ranunculus sceleratus* L. 'Celery-Leaved Crowfoot'
Ranunculus bulbosus L. 'Bulbous Buttercup'
crowefoot(C2) crakefote(C3) anglice crowefote(C12) anglice crofot or rammysfote(C26) ravynnisfet(C31)

Cornu Cervi
Plantago coronopus L. Buck's-Horn Plantain
more herbe yve, devills bitte(C21) morsus diaboli, pokyl-bytt [sic](C38)

Cornu Ircinum
Coronopus squamatus (Forsk.) Aschers. Swine-Cress
buckis horn, swynesgresse(C21)

Corona Regia [cortina r.]
Melilotus officinalis (L.) Pall. Melilot, Sweet Clover
Trifolium pratense L. Red Clover
mellilote, honisuke, red claver(A3)
gallice mellilote, anglice honisouc vel þe rede clavere(B2)
anglice honysokyls(C3) anglice honysouke vel rede clover(C6) mellilote, a thre-levyd grasse(C7) gallice mellilot, anglice þre-leved grasse(C12) rede clavyr, honysouke, tre-leved gras(C13) trifoli, honysukkill(C21) honisuge, anglice rede claver(C22)
Lit. N171

Corpore Epatica [corpore patica]
var. applied to *Agrimonia eupatoria* L., *Marchantia polymorpha* L., *Peltigera canina* Hoffm., and *Anemone hepatica* L.
anglice liverwort(B7)
anglice lyverwort(C6, C8)

Corrigiola [corigilla]

Polygonum aviculare L.	Knotgrass, Centinode, Swine's Grass, Sparrow Tongue
Lonicera periclymenum L.	Honeysuckle
Convolvulus arvensis L.	Bindweed
Glechoma hederacea L.	Ground Ivy

gallice cori(o)giole(A1) centinodie, knotede-gras, swynisgras(A3)
anglice heyhove, gallice ere terestre(B1) (certigeola) anglice sparewystunge(B1) gallice corigiole, anglice knotting-gras vel swynesgras(B2) wodebynde(B3)
gallice corrigiale, anglice clotgrasse [corr. cnotgrasse] vel swynesgrasse(C5) sparwistonge, onfortrede, quicchin(C13) centinody, knottegras, [s]wynegras(C22) anglice knottidgresse(C23)

Cortex Mandragore

?*Mandragora officinalis* L.	? Mandrake

canel(B7)
canel(C6)

Corulus [corilus]

Corylus avellana L.	Hazel

gallice coudre(A10)
anglice a hesyll vel newtre [corr. nut-tre?](C3) hasell(C28)

Costa

?*Plantago lanceolata* L.	? Ribwort

rybworte(C29)

Costus [costum, costrum]

Chrysanthemum balsamita L.	Costmary, Alecost, Balsamita

coste(A3)
gallice et anglice coste(B2) coste(B6) gallice et anglice cost(B7)
(c. maior) coste þe more (c. dulcis, minor) þe lasse coste(C1) costmarie(C2) anglice cooste(C6) anglice costmarye(C12) costmare(C26) þe bryn of whete*(C28) bran of whete*(C30)
Lit. André I, p.103 & II, p. 77

Cotida [for cotula fetida]
Anthemis cotula L. Stinking Chamomile, Mayweed
amarok, hondȝ-senemayn(C21)

Cotula Fetida
Anthemis cotula L. Stinking Chamomile, Mayweed
mai[th]e(B5)
þe mayþe vel camamyle(C6) gallice fumerol, anglice mathen(C16) amaroch, anglice maythe(C16) houndfenkel or maithe(C21) maies or hondz-senemayn(C21)
mousere*(C28)

Cotyledon [cotulidion]
[*Umbilicus rupestris* (Salisb.) Dandy]
voystel*(A11)

Crassula Aquatica [c. acuta]
Veronica beccabunga L. Brooklime
gallice faverole, anglice broclemke(B1)
broklemk(C28)

Crassula Maior
Sedum telephium L. Orpine
gallice orpin, anglice hoswort(A2) gallice crasse-geline vel orpine(A4) orpin(A7) gallice orpine, anglice aleswort(A8) gallice orpin(A9) gallice orpin, anglice haswort(A12)
anglice orpin(B2) orpyn(B3) gallice opie [sic], anglice haleswithe(B5) gallice orpyn, anglice halsour . . . [corr. halsuort](B9)
anglice orpyn(C6) orval(C19) orvayle(C21) gallice orpyn, madder*(C29) gallice orpyn, anglice waswort [corr. haswort?](C32) alswort, gallice orvale(C39)

Crassula Media
?
lyverwort(B3)
anglice lyverwort(C12)

Crassula Minor
Sedum acre L. Common Stonecrop
teste de soriz, stoncroppe(A3) gallice terce [sic] de soris, anglice stoncrop(A4) stoncrop(A7) anglice stancrop(A8)
teste de soris, anglice stancroppes(B2) stoncrop(B3)
anglice stanecroppe(C3) gallice test de soris, anglice stancrop(C6)

orivale, orpyn, stonhore(C7) stonore(C10) teste de surice, stoncrop(C13) souþistil*(C19) stonecrop or stonehore(C21) teste de fer [corr. soris], stoncroppe(C22) stonhore(C24) stonor(C27)

Cresso Aquaticus
Rorippa nasturtium-aquaticum (L.) Hayek — Watercress
watercress(C3) anglice watercresses or whit cresses(C26)

Cretanus Marinus [creta marina]
Eryngium maritimum L. — Sea Holly, Sea Hulver
Crithmum maritimum L. — Rock Samphire
watercretan, cretan of þe see, pety cretan, seevisc, watervisc, seehulfur, waterhulfur, pety hulfur(C9) (c. terrestris) hulfur-tre(C9) crete marine(C21) anglice wylde tesyl or see-holine growing by þe se(C26)
Lit. André I, pp.104f. & II, p. 78

Criola
Pastinaca sativa L. — Wild Parsnip
Sium sisarum L. — Skirret
skyrwytt(C7) anglice skyrewites(C19)

Crision [crison]
Galium aparine L. — Cleavers
clevere(B3)
clover*(C1) cliveren(C2) cliverern(C12)

Crisites
Achillea millefolium L. — Yarrow, Milfoil
millefoile, yargh(C21)

Crisocantis [crisacantis]
[*Hedera* L.]
methe*(A11)
osymund*(C21)

Crisolocanna [crisolicanna, grisolocanna, grisanacana]
Atriplex L. ssp. — Orache
melde(B1) medles(B3) anglice melde(B5)
mellers*, mespileres*(C2) mede or arrach(C3) anglice medlers*(C12) anglice melde(C20) wild arage(C21) meldis or wild arage(C21) arrage(C27) myld(C29)

Crispa Malva [cappa m.]
?*Geranium molle* L./*columbinum* L. ? 'Dove's-Foot Cranesbill'/ 'Long-Stalked Cranesbill'
?*Malva moschata* L. ? Musk Mallow
?*Malva crispa* L. ? 'Crisped Mallow'
anglice medehewort(A2) anglice mathwrt(A6) anglice madewort(A9)
anglice matwort(A12)
anglice macwort(B9)
matewort(C3) anglice crispe hocke(C20) matwort(C34)
Lit. Alph., p. 23, 1; Agnus Castus pp. 247f. see **Malva Crispa**

Crispina
?*Ribes uva-crispa* L. ? Gooseberry
crispine(C21)
Lit. cf. B III, 258 (crispinella)

Crista Galli
[*Salvia sclarea* L., *Rhinanthus crista-galli* L.]
thre-levedgres*(B4)
Lit. cf. André I, p.105 & II, pp. 78f.

Cristomelle
Glechoma hederacea L. Ground Ivy
ground-yvi, hors-yvy(C21)

Croalia
Allium vineale L. Crow Garlic
crogarlek(C9)

Crocus
Crocus sativus L. Saffron, Crocus
Carthamus tinctorius L. Safflower, Bastard Saffron
gallice saffrane(A1) safran(A3) gallice saffrane(A4) safran(A10) safran(B1) gallice et anglice safran(B2) saffron(B3) safron(B6) (campestris) wyld safron (almanicus) safron of Almayne(C1) (ortensis) safron de ort(C10) (ortensis) safforun d'orte(C11) safryn(C21) anglice saffrynsede or shalwort*(C26) saphryn(C31) (orientalis) safron de crai (campestris) wilde safron (almanicus) safron of Almayn(C38)

Cruciata Maior [herba c.m.]
?*Galium cruciata* (L.) Scop. ? Crosswort

warence or mader*(C6) more croswort(C21) þer ben .ii. of hem, herbe croyse(C27) (albus) cro[s]wort(C38)

Cruciata Minor
Galium cruciata (L.) Scop. Crosswort
herbe crose(B8)
croyser(C2) croys(C6) croyse(C7) herbe croyse(C10) herbe croise(C21) herbe poule*(C21) croswort(C38)

Cruciola [crusiola]
?
croyse, hulwort(A3)
croise, hillewort(C22)

Cubeba [cubebe]
Piper cubeba L. Cubeb
cubeb(C9) anglice quibibes(C12) quibibis(C26, C31)

Cucumer Agrestis
Ecballium elaterium A. Rich. 'Squirting Cucumber'
wilde nap(A7)
gallice cucumbre(B2)
(c. maior, lata) cucumbre (minor, rotunda) cucumbre (c. agrestis, silvaticus) wyld cucumbre(C1) carlok*(C21) kukumbre(C23)
Lit. André I, pp.106f. & II, p. 80; N172

Cucumeris
Bryonia dioica Jacq. Red and White Bryony
cucurbita agrestis, þe wylde vyne(C31)

Cucumerus
Cucumis sativus L. Cucumber
anglice horsement*(B1) werwette(B3)
Lit. B III, 149

Cucurbita
var. cucurbitaceae incl. *Bryonia dioica* Jacq., *Cucumis sativus* L.,*Citrullus colocynthis* (L.) Kuntze, *Lagenaria siceraria* (Mol.) Standl, *Lagenaria vulgaris* Ser.
gourde(A3)
gallice et anglice gourde(B2) nep(B3) wi[l]de nep(B5)

(longa) long gourd sulk (oblonga) gourd sulk (rotunda) gourd sulk (agrestis) wild nepe (Alexandri) gowrd Alisunder(C1) anglice gourde(C6) gowrde, potwort, þe grete snaylwort(C9) (agrestis) brionie, wild nepe(C10) gallice cucurbite, anglice gurd-sed(C16) gallice et anglice cucurbite .i. gou[r]de-sed(C18) anglice gurde-gras(C20) goorde or wyld nepe(C21) coule, couell(C24) col, stonel [corr. couel?](C27) Lit. André I, pp.107f. & II, p. 80

Culerage
Polygonum hydropiper L. — Smartweed, Water-Pepper, Culrage, Arsesmart
ersmert(B2)
anglice ersmynt [sic](C16) anglice erssmert(C18)

Culli Rabies
Polygonum hydropiper L. — Smartweed, Water-Pepper, Culrage, Arsesmart
anglice culerage(B7)
anglice culrage(C5) anglice cullarage(C6) culrage, arsmert, bitterwort(C21)

Cumille
?
chervell(C28)

Cuminum [ciminum]
Cuminum cyminum L. — Cumin
gallice cumyne(A1) comin(A3) cumyn(A4)
comyn(B1) gallice et anglice comyn(B2) comyn(B5)
comyn(C1)

Cuminum Agreste
Lagoecia cuminoides L. — Wild Cumin
wyld comyn(C1) wylde comyn(C21)

Cuminum Dulce
Pimpinella anisum L. — Anise
anglice aneyse(C12) gallice et anglice anneys(C16) anyse(C21)

Cuminum Ethiopicum
Nigella damascena L. — 'Love-in-a-Mist'
Nigella sativa L. — 'Black Cumin'

anglice cokkyl(C1) kokkel(C12) cokkill(C21)
Lit. cf. André I, p.108 & II, p. 81

Cuminum Rusticorum [ciminum r.]
Seseli tortuosum L. Sermountain
Siler Montanum Crantz Sermountain
sirmontayne(B3)
surmuntayne(C6) anglice syrmontayne(C12) sermontayne(C36)
Lit. cf. André I, p.108 & II, p. 81

Cuscuta [cuscute, cuscutis, cucuscata, cuscutus]
Cuscuta europaea L. 'Large Dodder'
Cuscuta epilinum Weihe 'Flax Dodder'
Cuscuta epithymum L. 'Common Dodder'
une herbe ke crest entre line, anglice dodre(A1) doder(A3) dodre(A4)
anglice dodre vel doder(A5) anglice dodire, berycluan(A6) doder(A7)
anglice doder(A9, A12) anglice deder(A10)
anglice doder(B2) dodire(B4) anglice doder(B5) doder(B6)
doder(C2) anglice dodyr(C3) dedere(C24) doedore(C37)

Custos Ortorum/Orti
?*Leonurus cardiaca* L. ? Motherwort
?*Ricinus communis* L. ? Castor Oil Plant
moderwort(B3)
clyþe*(C7)
Lit. N172

Dactilium
?
wodclete(C16)
Lit. cf. B III, 266

Dactilosa [dactilosus]
Paeonia L. ssp. esp. *corallina* Retz
pioyne(B3)
pynot-tre*(C9)
Lit. André I, p.115 & II, p. 86

Dactilus [dactalus]
Phoenix dactylifera L. Date
date(A3)
gallice et anglice date(B2)
date(C2) date-tre(C9)

Dactulorum
Rosmarinus officinalis L. Rosemary
nedilwort(C16)

Dactulus (Acetosus)
Tamarindus indica L. Tamarind
tamarindes, anglice sorel*(C10) tamaryndes(C27)

Dala
Rumex L. esp. *R. aquaticus* L. and *R. sanguineus* L. 'Red(-veined) Dock'
gallice parele, anglice rede docke(C6)

Damphinis [dampnus]
Laurus nobilis L. Laurel
Daphne laureola L./*mezereum* L. Spurge Laurel/Mezereon
anglice lorer(C12)
Lit. cf. André I, p.116 (daphnitis) & II, p. 87

Dampnides
Laurus nobilis L. Laurel
Daphne laureola L./*mezereum* L. Spurge Laurel/Mezereon
lorel-tre(C21, C27)
Lit. cf. André I, p.116 (daphnoides) & II, p. 87; N172

Dantes
Rosmarinus officinalis L. Rosemary
gallice rosmarine(B2)

Dardamon
?*Gentiana* L. esp. *G. amarella* (L.) Börner? 'Baldmoney', Gentian
baldmony(C21)

Dascripta [datrista]
'Epatica' ?
lyverwort(C10, C27) liverwort(C21)

Daucus [d. creticus, d. asininus]
Daucus carota L. Wild Carrot
Pastinaca sativa L. Wild Parsnip
?*Athamanta cretensis* L. ? Cretan Carrot
gallice dauk(A1)

(creticus) anglice walmore(B1) gallice dauk, pastinak chaumpetre(B2) dauke(B3) dawke(B8)
(creticus, agrestis minor) dauke (asininus, agrestis maior) bridesnest, bolwed*(C1) anglice wylde skyrwyȝt(C6) dauke, copwort*(C7) (creticus) dent de lyon*(C10) walmore vel muse-nest(C20) (asininus) anglice kex*(C20) (creticus, minor) lesse dauke (azininus, maior) birdisnest or more dauk(C21) ditandre*(C23) dauke, pastinat champetre(C23) dent de lyon*(C27) walmore(C28) (creticus) dauk or dawk or brydnest(C31) (asininus) bryddysneste (creticus) lesse dauk(C32) (creticus) skirwittis(C35) kexis*(C39)
Lit. B III, 249; N173

Dedagantrix [degargati]
Rosa rubiginosa L. Eglantine, Briar Rose
Rosa canina L. Dog Rose
eglenter, anglice hepe-brimble(B1)
gallice engleter, anglice hepe-brembul(C6) hep-bremble(C28)

Degalia
Cynoglossum officinale L. Hound's-Tongue
lang de chen, anglice hundestong(B1)
gallice lange de chien, anglice houndestong(C6)
Lit. DMLBS degalea

Delphinion [delphinon]
Delphinium ambiguum L. Larkspur
folispir(B1) fouloȝwyse(B1)
fulsp(t)ir(C28) anglice fouleswyse(C41)
Lit. B III, 100

Demechia [demachian]
Daemonorops draco Blume ex Schult. Dragon's Blood (resin)
dragons-blod, sangdragon(C9) anglice dragans-blode(C26)
Lit. Alph. App. p. 211 (damecham)

Dendrolibanum
Rosmarinus officinalis L. Rosemary
anglice rosemaryne(C3) rosmaryne(C10) rosemarine(C21)

Dens Caballinus
Hyoscyamus niger L. Henbane
henbane(C21)

Dens Canis
?*Agropyron caninum* (L.) Beauv. ? 'Bearded Couch-Grass'
hondestoþ(C32)

Dens Equi [d. equus, d. equinus]
?*Peucedanum officinale* L. ? Hog's Fennel, Sulphur-Weed
?*Trifolium ochroleucon* Huds. ? 'Sulphur Clover'
?*Cyperus longus* L. ? Galingale
?*Lotus corniculatus* L. ? Birdsfoot Trefoil
?*Lathyrus tuberosus* L. ? 'Earth-Nut Pea'
?*Vicia hirsuta* (L.) S.F. Gray ? Hairy Tare
horstoþe(C1) horstoth, lytel snaylwort, sulfurat, pekkokystayl, wild pese(C9) (aquatica) water-horstoth(C9) horstoth(C21) ciperi, horsistoþ(C35) anglice tyntare(C41)
Lit. Alph., p. 49, 8, N173

Dens Leonis
Taraxacum officinale Weber Dandelion
dent de lion(A3) gallice dent de leon(A4) gallice dent leoun(A6) gallice dent de lioun, anglice liones-totȝ(B2) dent de lyoun(B3) box*(B5) (maior) monkeshed, dent lyon (d.l., dens canis) houndystoþe, lasse dent lyon(C1) daundelyon(C2) anglice preste-cron(C3) gallice dent de lioun, anglice carterus-sauge vel lyonus-toþe, bytterworte(C6) dent de lyon, bytirwort(C7) gallice dent de lyon, anglice carteres-sage(C8) gallice dent de lyon, anglice lyontothe(C16) preste-croune(C29) prestis-croune(C35) anglice carter-sauge(C36) monkyshead(C38) gallice dent lyon, anglice doloroune(C41) honstonge*[corr. lionstoþe](C43)
Lit. Alph., p. 49, 5 (doleroune)

Dens Lupi
?
dent de low(B3)
wolvestoþe(C1) wulvystoth(C21) wolvustoþ(C32) wolfistoþe(C38)

Dent de Lyoun
?
dedman flour(C14, C15)

Dentaria
?*Chrysanthemum parthenium* (L.) Bernh. ? Feverfew
?*Anacyclus pyrethrum* DC. ? Pellitory

poseire(B3)
Lit. SB p.17 (dentaria i. piretrum); André I, p.117 & II, p. 88

Deo Gratias
?
Dieu-grace(C22)

Detonica
?*Dictamnus albus* L. — ? Bastard-/White Dittany
detoyn(B8)

Diaforetica
Althaea officinalis L. — Marsh Mallow
anglice wymauve(A5)

Dibra
Urtica L. ssp. — Nettle
netell(C7) anglice netell(C12)

Digalea
?*Ranunculus sceleratus* L. — 'Celery-Leaved Crowfoot'
cloftonge(C37)

Digitus Veneris
?*Antirrhinum orontium* L. — ? Snapdragon
?*Linaria* L. ssp. — ? Toadflax
Cynoglossum officinale L. — Hound's-Tongue
houndesheed(C21) anglice stychewort* or hondestong(C26)

Dilla
Rumex L. ssp. — Dock
paradella, gallice parele, anglice doke(B1)
rede dok(C28)
Lit. B III, 71

Dionisia
Calendula officinalis L. — Marigold
þe flur of rodene(C10) anglice roden(C12) þe flour of roden(C24)
Lit. Alph., p. 50, 6. cf. André I, p.119 (Dionysias) & II, p. 89

Dipsacus
[*Dipsacus fullonum* L./*sylvestris* Huds. (Wild) Teasel]
conf. w. diptamnus
ditonge*(C21)

Diptania Silvestris
?*Polygonatum multiflorum* (L.) All. ? Solomon's Seal
?*Hypericum androsaemum* L. ? Tutsan
sigillum Salamonis, Salamonsele, tutelayne [corr. tutesayne?](B4)
cuculeyne, totesan(C7) sigillum Salomonis, tuteceleyne(C12)
Lit. Alph., p. 51, 4 (Diptannum . . . folia habet lanosa et super omnem fruticem velut lana)

Diptannus [dactanus, diptannum, diptamum, diptania, ditania, detanus]
Origanum dictamnus L. Dittany of Crete
Dictamnus albus L. Bastard-/White Dittany
conf. w. *Lepidium latifolium* L. Dittander
gallice ditayne, anglice ditaundir(A1) gallice et anglice dytoyne(A2) ditandre(A3) gallice et anglice ditaundre(A6) ditaine(A8) anglice dytayne(A9) gallice et anglice ditanie(A12)
ditayn(B1) gallice et anglice ditaigne(B2) dytayne(B3) detayne(B4) gallice dytayne(B5) dytayne(B6)
dytayne, dytaundyr(C1) dytayne(C2) ditandre(C22) diptayne(C23) peperwort(C32)

Diptannus Minor
?*Chrysanthemum balsamita* L. ? Costmary
coste(C21)

Diptoli
Polypodium vulgare L. 'Polypody'
Thelypteris dryopteris (L.) Slosson Oak Fern
polipodi(C21)

Discopolidis
Bryonia dioica Jacq. Bryony
wild nepe(C21)
Lit. André I, p.119 (discopela)

Ditropia
Calendula officinalis L. Marigold
anglice marigolde(C3)

Dominella
?*Carum carvi* L. ? Caraway
carwi(C21)

Domitrix
Agrimonia eupatoria L. Agrimony
egrimonie(C21)
Lit. André I, p.120 (domitria) & II, p. 90

Dormentille
Urtica dioica L./*urens* L. Stinging-/Small-Nettle
gallice orte grech, anglice blyndnettyll(C16)

Dormonilla [dormenella]
?*Lactuca sativa* L. ? Garden Lettuce
herbe dormant, slepyngwurd(B1)
schepwort [corr. slepwort](C14)

Dragantea [draguncia]
Polygonum bistorta L. Bistort, Adderwort
Dracunculus vulgaris Schott Dragon Arum
serpentine, edderwert(A3)
gallice dragaunz, anglice nedderwort(B1) dragaunce(B3) dragaunce, serpentyne(B4)
(d. masculina) dragons, edderwort (d. femina) dragons, eddyswort(C1)
anglice dragance(C6) serpentary(C42)

Dracontea
Dracunculus vulgaris Schott Dragon Arum
dragance(A11)

Dragancia
?*Sinapis arvensis* L. ? Charlock, 'Wild Mustard'
?*Raphanus raphanistrum* L. ? Wild Radish, White Charlock
cherlok(C11) cherlonge [sic] (C24) cherloke(C27) stanchwort*(C28)

Driopes
Thelypteris dryopteris (L.) Slosson Oak Fern
gallice polipodie, anglice hocfern(B1)
okeferne(C6) overfarn(C28)
Lit. André, p.12 (dryopteris) & II, p. 91; N174

Drissitis
Teucrium chamaedrys L. 'Wall Germander'
germaundre(C21)
Lit. Alph. App. p. 211 (disitis vel drisitis) ; André I, p.121 (drys) & II, p. 91

Droptori
[*Polypodium vulgare* L. 'Polypody']
the flour of ruddene*(C21)
Lit. André I, p.121 (dryopteris) & II, p. 91

Drumphana [drumfama]
var. *cucurbitaceae* incl. *Lagenaria vulgaris* Ser. 'Gourd'
anglice gordine(B3)
gourden(C2) anglice gorden(C12)

Duma
Ranunculus flammula L. Lesser Spearwort
sperwort(C16)

Ebulus [ebulum]
Sambucus ebulus L. Danewort, Dwarf Elder, 'Walewort'
gallice ebil, anglice walewort(A1) gallice eble, anglice walewort(A2) eble, walwort(A3) gallice eble, anglice wallewort(A4) gallice eble, anglice walwrt(A6) walewort(A7) gallice eble, anglice wa[l]ewort(A8) gallice eble, anglice wa[l]wort(A9) walewort(A11) gallice eble, anglice walwort(A12)
gallice eble, anglice walwort(B1) gallice eble, anglice walewort(B2) eble, walwort(B3) eble, walworte(B4) gallice eble, anglice walewrt(B5) walwort(B6)
wallewort (maius) þe more (minus) þe lasse walwort(C1) eble, anglice walworte(C3) lyþewort(C7) (niger) lyþerwort(C7) eble, lythewort or walwort(C10) gallice ebul(C16) (minor) lesse walwort(C21) eble, liþewort or walwort(C31)
Lit. B II, 79f & III, 163

Edoc [eudoke, edokke]
Nuphar Sm. ssp. Water-Lily
waterlilie(B2) anglice waturlily(B7)
anglice waterlyly(C1) anglice waturlyly(C6) anglice waterlelye(C22)

Eglenter [eglentera]
Rosa rubiginosa L. Eglantine, Briar Rose
Rosa canina L. Dog Rose
anglice huptre(B7)
anglice heoptre(C6) anglice eglentyne(C12) hupbrere(C21) hupebrere(C36)

Elaborites [eloborica]
Centaurium erythraea Rafn Common Centaury
Blackstonia perfoliata (L.) Huds. Yellow-Wort
?*Artemisia vulgaris* L. ? Mugwort
(emeracleus, emipiteus) lesse centori, cristeleddre(C21) anglice fetherfew or less centory or lesse mogwort(C26)
Lit. André I, p.125 (elleborites) & II, p. 94

Elacterides [elacteride, elachiodos]
Euphorbia lathyrus L. Caper Spurge
juce of cat(r)epus .i. juce of spurge(C10) the juse of catep(r)us .i. spurge(C21)
Lit. Alph., p. 93, 12; N174f

Elacterium [electericum, electuarium, electarium, electorium]
Euphorbia lathyrus L. Caper Spurge
Ecballium elaterium A. Rich. Squirting Cucumber
spourge(B2) anglice sporge(B7)
spurge(C1) anglice spurge(C6) þe juce of wilde cucumbre .i. wilde papie(C10) anglice moderwort*(C12) anglice spourge(C18) anglice moderwort*(C20) sporge(C21) ye juse of wild papy(C21) moderwort*(C23) þe juse of wilde cucumbre, wilde popy(C24)
Lit. André I, p.124 (elaterium) & II, p. 93; N175

Electarus
?*Senecio jacobaea* L. ? Ragwort
regedewort(A2)

Electra [electrum]
Saponaria officinalis L. Soapwort
Dipsacus fullonum L. Teasel
crousope(C28) anglice tesel(C36)

Electrum Nigrum [elostrum n.]
?*Lupinus* L. ssp. ? Lupin
?*Inula helenium* L. ? Elecampane

?*Scrophularia nodosa* L. ? Figwort
scabgras(A3)
scabbewort(B5)
scabbegras(C22)
Lit. B III, 79

Elempnium [elempruum, elempinum]
Inula helenium L. Elecampane, Horseheal
horselne(C10) horshelne(C21, C27) horself(C24) anglice broke-medewort or helena campane; ther ben .ii. maners of enula campana, on with a yalow floure and another with a whyte, but the beter hath a yalow flour(C26)

Elena
Inula helenium L. Elecampane
alne, grotwort [= gretwort](B8)

Elephobotus
[?*Peucedanum ostruthium* (L.) Koch ? Master-Wort]
[?*Pastinaca sativa* L. ? Wild Carrot]
anglice hindesberie*(B1)
Lit. André I, p.124 (elaphoboscon) & II, p. 93

Elerium [elirium; for esbrium?]
Salvia L. ssp. Sage
gallice [sic] et anglice floures of sauge(C16)
Lit. see **Esbrium**

Elicrodanum
Lonicera periclymenum L. Honeysuckle
Convolvulus arvensis L. Bindweed
wodebynd(C21)
Lit. Alph., p. 55, 21 (elitrodamum)

Elidrius
Populus L. ssp. Poplar
þe popil tre(C9)
Lit. Alph., p. 55, 15 (elitridium)

Elifagus
Salvia officinalis L. Sage
sauge(C9)

Elixemum [elixemium]
Lilium L. ssp. — Lily
lyly(C21)
Lit. Alph., p. 55, 23

Elixis [elix]
Arctium lappa L. — Burdock
anglice clote(C10) cloter(C21, C27)
Lit. see **Barba Elixis**

Elleborus Mortificativus
Helleborus L. ssp. — Hellebore, var. poisonous plants?
muchewort(A3)

Elleborus (Albus) [elibarus, elleborum]
Helleborus foetidus L./*viridis* L. — Bear's-Foot
Veratrum album L. — White Hellebore
elebre, thathau(A3) anglice attreloberye*(A6)
un maner de elebre, gallice elebre blanc, anglice michelwort(B2) elebire, bischopworte(B4) elebyre whit and blak(B4)
whyt, a lasse ebelur [sic](C1) bysshoppeswort(C2) ellebore whit(C2) (albus et niger) langworth et clovetong(C3) cloftyng vel mykylwort, bishoppsworte or bitterwort(C6) whit ellerne*[sic](C10) anglice seterwort(C12) mykelwort, dwale(C15) gallice elebre blank, anglice mechilwort(C16) bischopiswort(C19) anglice longwort(C20) lungwort(C21) elebre, mugwort(C22) anglice muchewrt(C23) white elebre(C27) pé de bbere[!](C28) peleter of Spayn*(C33) qwhyt peletter of Spayne*(C34) gallice syfonye*(C39)
Lit. Alph., p.189 (vellatrum); B III, 164 (lungenwort) & 239 ([th]ung)

Elleborus Niger [eliborus n., elleborum n.]
Helleborus niger L. — Black Hellebore
Helleborus foetidus L./*viridis* L. — Bear's-Foot
and var. poisonous plants
suffonie, cloftunge(A3) anglice mogwrt(A6) anglice longwort(A10) anglice tedeberie vel bissopwort(B1) gallice suffoire[sic], anglice cleftonge(B2) lungworte(B4)
cloveþonge(C1) elebore blak(C2) gallice suffoyne, anglice todebery, longwort(C6) gallice suffoyr[sic], anglice cloftonge vel mykelwort(C9) clovetonge, blak ellerne(C10) s[e]terworte or longwort(C12) longwort(C19) anglice mikeword(C20) blak elebre, cloftung, pe de lyon ye more(C21) suffroire[sic], dostunghe [corr. cloftunghe](C23)

clovetong, blak elebour(C24) clovetong, blak elebyr(C27) se(n)trewort(C28) elebyr, loungworte, lowungeworte(C33) clovetonge, blak elebre(C40) clovetong or pedelyon(C41)
Lit. B III, 164 & 239

Ematicos [emaideos]
Rubus L. ssp. — Bramble
brere, bramelis(C1) brere, bremmyll(C38)

Emeracleus
Centaurium erythraea Rafn — Common Centaury
more centory or erthgalle(C21)
Lit. André I, p.125 (emericos); see **Elaborites**

Empus [vel emtus]
?*Lonicera periclymenum* L. — ? Honeysuckle
?*Convolvulus arvensis* L. — ? Bindweed
anglice wodbynde(C12)

Emtis
?*Arctium lappa* L. — ? Burdock
bur, clote(C21)

Endivia [endiva]
Sonchus oleraceus L. — Milk- or Sow-Thistle
Lactuca serriola L./*virosa* L. — 'Prickly Lettuce'/'Bitter Lettuce'
Cichorium intybus L./*endivia* L. — Chicory, Wild Succory/Endive
scariele, souþestyl(A3) gallice endive(A4)
gallice et anglice endive(B1) gallice scariole, anglice soutþestel(B2) endyve(B3) endive(B5)
endyve, wyld letuse, horsthystyl(C1) gallice [. . .] anglice þistel(C8) tame endif, wilde endif(C9) scariole, southustil(C16) anglice þoweþistel(C20) endive, horthistill(C21)

Enucleatar [enucleater]
Rosa rubiginosa L. — Eglantine, Briar Rose
Rosa canina L. — Dog Rose
gallice eruce vel eglenter, anglice hepe-tre(B2)
gallice eruce vel eglenter, anglice hupis, wodegarid(C16) gallice eruce vel eglenter, anglice hupe-tre(C18)

Enula Campana [elena campana]
Inula helenium L. Elecampane, Scabwort, Horseheal
gallice auné, anglice horshelne(A1) ané, horselle(A3) gallice auné, anglice horshelne(A4) horsheln(A8)
gallice auné, anglice horshelne(B1) gallice yauné, anglice horshelne(B2) horselne(B3) horshelne(B4) horselne(B5) anglice horshelm(B9)
horselrun(C1) anglice horsehelme(C3) horshellern(C6) auris equina, horsere(C9) horshelre(C15) anglice horshelne, scabwort(C16) horselene more(C20) horsehele(C22) horshelme(C28) horssalne(C30) schabwort, horse-elerne(C31) gallice auné anglice hersell[sic](C39) scabbewort(C39) anglice elen campan(C41) romani vocant herb grosse, anglici scabwort(C43)

Epalsion
?*Potentilla reptans* L. ? Cinquefoil
five-leved gress(C21)

Epatica [eupatica, epatik, epaticum]
?*Anemone hepatica* L. ? 'Noble Liverwort'
?*Marchantia polymorpha* L. ? Liverwort
gallice coper, anglice liverwort(A1) anglice liverwort(A2) eupatik, coper, livyrwort(A3) gallice epatik vel coper, anglice livrewort(A4) anglice lyverwrt(A6) anglice lewrewort(A8)
gallice epatik, anglice liverswort(B2) lyverwort(B3) leverworte(B4) anglice liverwrt(B5) lyverwort(B6) gallice epatik, anglice liverwort(B7)
liverwort(C1) gallice copire, liverwort anglice(C2) anglice lyverworth(C3) gallice eupatike, anglice liverwort vel wy[l]de sound(C5) (aquatica et montana) caper [corr. coper], lyverwort(C6) coupere, lyverwort(C10) eupatyk, coper, anglice lyverewort(C22) iecurina, anglice lyverwort(C26) coupere, liverworte(C27) liverwort vel wil[d] fond [corr. sond?](C36)

Epaucius
some kind of thistle
anglice brode-thistil(B1)
brode-þistel(C28)
Lit. B III, 37

Epesia
['Artemesia']
elisticus, red madre*(C21)
Lit. André I, pp.126 (Ephesia) & 125 (elestitis) & II, p. 95

Epicurium [epiturium]
?*Sedum telephium* L. ? Orpine
anglice halswort .i. orpin(B1)
anglice orpyn(C6) halswert(C28)
Lit. B III, 121f

Epithymum
Cuscuta epithymum L. Thyme Dodder
flour of time(C10) anglice tyme(C12) the flour of tyme(C21)

Eptaphiron
Plantago major L. Plantain, Waybread
(eptafilos, eptaplinos, eptapleurion: plantago, C1) planteyn(C21)
Lit. B III, 107f; André I, p.160 (heptapleuron) & II, p.120

Eptome
Cynoglossum officinale L. Hound's-Tongue Fern
houndstung(C21)

Erbus [ervus, orbus, orobus]
Vicia L. ssp. Vetch, Tare
tar-vecche, fech(C9) anglice mouseher(C12) ervus dicitur harmel*(C13)

Erifion [eufion, eriphion]
'epatica' [error?]
anglice litilwort [corr. lithwort](B1)
liverwort(C1,C21) lithewort(C28) litkurt [corr. litwurt](C37)
Lit. B III, 163f

Erimon
?*Euphorbia* L. ssp. ? Spurge
titimallus, healbrede(C37)

Eritridanum [eritudanum, entidanum, etrydanus, etrydana, entridanum]
Rubia tinctorum L. Madder
þe grete mader(C9) gallice warance, anglice mader(C12) madir(C21)
anglice madre(C26)
Lit. Alph., p. 59, 9 (eritridanum vel eritrodanum); André I, pp.127 (ereuthodanum) & 129 (erythrodanos) & II, pp. 96 & 98

Ermion [ermeon]
Lilium L. ssp. Lily

gallice liz, anglice lilie(B1)
anglice lylye(C6) lyly(C28)

Ermocalles
?*Dipsacus sylvestris* Huds. ? Wild Teasel
?*Sonchus oleraceus* L. ? Sow-Thistle
haretasel(B1)
wilde tasell(C28)
Lit. cf. André I, p.159 (hemerocalles) & II, p.119

Ermula
?*Mercurialis perennis* L. Dog's Mercury
?*Chenopodium bonus-henricus* L. Mercury, Allgood
mercury, papwort or more smerwort(C21)

Erpina
?*Digitalis purpurea* L. ? Foxglove
?*Convallaria majalis* L. ? Lily-of-the-Valley
foxglove, glovewort(C1) foxeglove(C32) foxglove(C38) foxclove(C41)
Lit. Agnus Castus p. 222

Eruca [euruca]
Eruca sativa Mill. Garden Rocket, White Rocket
Sium L. ssp. Skirret
Raphanus raphanistrum L. Wild Radish, White Charlok, Runch
Sinapis arvensis L. Charlok
gallice eruk, anglice wyte-pepir(A1) anglice skyrwit(A2) eurik, wyt-peppir(A3) gallice eruk(A4) squirwit anglice, trummor anglice(A5) anglice skyrewyt(A6) anglice skirewis(A8) anglice scirewit(A9) anglice schyrwyt(A12)
gallice eruc, anglice ȝwit-peper(B2) erukes, skirwhittes(B4) anglice white pepir(B7) anglice skirwit(B9)
(maior) ruge .i. piper, allium secundum alios cherloc (minor) rulkull(C1) schirwort(C7) erukes, anglice skyrwyttes(C12) (maior) ozimunde*, (minor) rukkill(C21) anglice white peper or whit musterdsede or ketelock [corr. kerelock] or skyrwhyt(C26) warlock of he . . .(C29) ruge (minor) kokyll [corr. rokyll](C38)

Erusa
?*Arctium lappa* L. ? Burdock
clote(B3)
anglice clote(C12)

Erustix [erustrix]
Rubus L. ssp. Bramble
gallice rounce, anglice brimbil(B1)
brambell(C28)
Lit. B III, 37f; André p.128 (eruscum)

Ervicella
Sedum acre L. Stonecrop
anglice stonhore(C12)

Esbrium [elbrium]
Salvia officinalis L. Sage
flur de sauge(A3)
gallice flours de sauge(B2)
þe flowr of sage(C10) anglice tame sage(C26)
Lit. Alph., p. 58, 18; MS ccxliii ('esbrium .i. salvia vel flos eius, sed falso; immo est sansucus vel majorana'); N176

Escula
Mespilus germanica L. Medlar
openers, medelers(C31)

Esophi
?*Potentilla reptans* L. ? Cinquefoil
five-leved gresse(C21)

Esula
Euphorbia L. ssp. Spurge
esle(A3)
esule(C2), on of þe spurge(C9) anglice esule(C12) esle(C22) esul(C23)

Etinta [etincha, etinca]
?*Pastinaca sativa* L. ? Wild Parsnip
?*Sium* L. ssp. ? Water Parsnip
anglice skirwhit(B7)
anglice skyrwhyth(C6)

Eubucone
Sambucus ebulus L. Danewort, Dwarf Elder
þe more walwort(C1)
Lit. André I, p.123 (ebucone)

Eufordaphyn
?*Cinnamomum zeylanicum* Blume ? Cinnamon
?*Cinnamomum cassia* (L.) Blume ? Bastard Cinnamon, Cassia
canell(C27)

Eupatorium [epatorium]
Teucrium scorodonia L. Wood Sage, 'Sage-Leaved Germander', Hind Heal
gallice eupatorie, anglice wylde sage .i. hyndehale(A1) eupatorie, wylde sauge(A3) sauge sauvage(A7)
gallice eupa[to]rie, anglice wilde sauge(B2)
gallice eupatorie, anglice wylde sauge(C6) anglice eupatorye(C8) anglice wodesauge, gallice ambrosie(C20) houndisbery*(C21) hyndehale(C24) smalewort*(C34) smerworte*(C37)
Lit. N177

Euphorbium
Euphorbia L. Spurge
[= gummi/herba] gallice euforbe(A5)
clarettum, anglice clarry*(C16) clarrey*(C21) (conf. w. euphrasia); gomme of þe blak thorn(C28) clareye*(C32)

Euphrasia [eufrasea]
Euphrasia L. ssp. Eyebright
anglice eufrasie(A6)
gallice et anglice eufrase(B1) eufrase(B3) anglice eufrase(B7)
eufrase(C1, C2) gallice et anglice eufrasye .i. ibryȝte(C6) eufras(C36) birdistonge* vel eufras(C39)

Evictus [?]
?*Crocus sativus* L. ? Saffron
anglice saferon(C12)

Eviscus
Althaea officinalis L. Marsh Mallow
wymalve, anglice hocke(C10) holihok(C21) þe wyld malwe(C27)

Exifion
Iris L. ssp. esp *foetidissima* L. Gladden, Flag
gladyn or bladyn(C21)
Lit. André I, p. 339 (xiphium) & II, 278. cf. B III, 97f

Faba
Vicia faba L. Bean
(f. dura) feve, bene (f. virides) feve, bene(A3)
gallice feves, anglice benes(B2)
(f. terrestris, communis) fuly-bene (f. Indie) ben of Ynde(C1)

Faba Lupina
Atropa bella-donna L. Dwale, Deadly Nightshade
dwale(C9)

Fabaria [fabercula, faberia]
Veronica beccabunga L. Brooklime
gallice faverole, anglice lemek(A1) anglice lymyke(A2) faverole, lemeke(A3) gallice faverole, anglice leymke(A6) gallice[sic] lameche vel lemche(A8) gallice faverol, anglice lemke(A9, A12)
gallice faverole, anglice broklemke(B2) lemeke(B3) feveroyle(B4) gallice feveroles, anglice lemeke(B5) gallice faveroyl, anglice lemp(B9)
(f. minor, aquatica) more lenk (media, aquatica) myddyl lenk (minor, agrestis) feld-lenk(C1) broklenke, gallice faverole(C2) faverole, lemke or brokelempe(C7) waterbene(C9) (maior) brokelenyk, faverole (minor) les brokelenyk(C21) faverole, lem(C24) broke-lily(C28) (minor) lasse broklemke(C32) (maior) þe more lenk(C38) (minor) lesse brokelemke, hokke-levys* [corr. brokke-levys] (aquatica) brokmodurwort or broklemk(C41) broklevys(C42)

Facasnion
Iris L. ssp. Gladiolus, Gladden
gladene(C21)
Lit. André I, p. 247 (phasganion) & II, p.196; Alph., p. 62, 10 (fagasnion)

Faciolus [facellus]
Phaseolus vulgaris L. Kidney Bean
(f. albus, maior) more faciole, fasole (rubeus, minor) faciole þe lasse(C1)
herba beati Petri, primula veris, prymerose*(C7) (albus) ffasell(C21)
(albus) facyol(C38)

Fagus
Fagus L. Beech
anglice beche(B7)
byrche*(C28)

Falgeria [feugera, flageria, feraria]
var. *polypodiaceae* Fern

Osmunda regalis L. Royal Fern
feugere, verne(A3)
f. vel hosmunda, verveyn*(B1)
braken(C15) feuger, anglice ferne(C22) dichfern(C28)

Famellus
Lysimachia L. ssp. (Yellow) Loosestrife
herbe Cristofre or jaunnez(C21)

Farfar
Tussilago farfara L. Coltsfoot
anglice horshove(B1)
horshove(C6, C28)
Lit. André I, p.133 & II, p.102

Fasolaria
?*Chelidonium majus* L. ? Greater Celandine
?*Glaucium corniculatum* (L.) Rudolph/ *flavum* Crantz ? 'Red/Yellow Horned-Poppy'
celidoyne(B3)

Fassia
Veronica beccabunga L. Brooklime
gallice faverole, anglice lenk(B1)

Faverola
Veronica beccabunga L. Brooklime
brokelemke(C19)

Favida
Veronica beccabunga L. Brooklime
favede, lymeke(A2)
lemke(C29)
Lit. cf. B III, 136f

Febrifuga [febrifigia, februfugia, febrefuga]
Chrysanthemum parthenium (L.) Bernh. Feverfew
Centaurium erythraea Rafn Common Centaury
gallice feutherefoy, anglice holywort(A1) veþerfoy, erþegalle(A3)
anglice fefyrfoy(A10)
feverfoile, anglice fetherfoie(B1) gallice feverefoy(B2) fetherfewe(B4) everfern*(B5) fevirfoie(B5) anglice feverfew(B7) gallice[sic] feþerfu, anglice aldwort(B9)

fetherfoye(C2) anglice fedyrfewe(C3) anglice feþerfoye(C6) fetherfoy, witwort(C7) centaurea minor, anglice, gallice fetherfoye(C16) gallice erigalle [corr. erþgale](C19) fedirfew(C21) vethervoy, anglice erthegalle(C22) feyesfoye(C27) fedrefoy(C28) holywort, fetherfoy(C32) feverfew or veturfoy(C41) fethirfoy, hersgall(C42)

Fel Terre
Centaurium erythraea Rafn — Centaury, Christ's Ladder, Feverwort
centorie, cristis laddre(A3)
centorie(B3, B5)
anglice feltworth*(C6) felwort(C8) more centori(C21) anglice centory, cristesladdre(C22) feldwort*(C36)

Feltrum Terre [feltrix t., filtrum t., felix t.]
conf. of fel terre 'feltrike' q.v. and filtrum 'feltwort' [*Verbascum thapsus* L., Great Mullein]
anglice feltrike(A6) anglice feltwort(A8) anglice feltrike(A9) anglice feltryke(A12)
feldwort(B3) anglice feltwort(B5) anglice feltrik vel feltwort(B9)
anglice felt[r]yke(C3) feldworth(C15) feldwort, erthgall(C32)
fellewort(C37) feltrewort(C41)

Fendes
Anthemis cotula L. — Stinking Chamomile, Mayweed
pes columbinus, herba fetida, anglice maythworth(C3)
Lit. Alph. App. p. 214 (fenda)

Feniculus Porcinus [feniculum porcinum]
Peucedanum officinale L. — Hog's Fennel
swynysfynel, wormesed(C1) anglice swynefenkell(C3)
houndesfenel*(C10) swyneȝthistell*, wormseede(C21) swynsgrece(C28)
swynesfenel, wormsede(C32)

Feniculus [feniculum]
Foeniculum vulgare Mill. — Fennel, Fenkel
gallice fenycle, anglice fenekyl(A1) fenyl(A3) gallice feneyl(A4)
gallice fenel, anglice fenekele(B2) fenkell(B6)
(f. albus) whyt fenel (f. rubeus) red fenel(C1) fenel(C2) anglice fynkell(C3) anglice fenell(C6) anglice fenel vel fenekel(C18)
fynkill(C24) ffenkle(C28) woode-fenel(C31)

Feniculum Sancte Marie [feniculus s.M.]
?*Chrysanthemum vulgare* (L.) Bernh. ? Tansy
tanasi(B3) anglice tanesay(B9)
anglice tansye(C12) maidwede(C21) camamille*(C35)
Lit. cf. Alph., p. 63, 25; SB 21. ('tadefenel')

Fenugrecum [funugrecum, fenigrecum]
Trigonella foenum-graecum L. Fenugreek
fenugrec(A3) gallice fenugrek(A4)
gallice et anglice fenugrec(B2) fenugrek(B3) fenugreke(B4)
funugreke (f. agreste) wyld fingreke(C1) fenegreke(C2) fenucrek(C22)

Ferula
?*Plantago lanceolata* L. ? Ribwort
ribwort(C21)
Lit. cf. Alph., p. 64, 1 (and note)
C3 = minuta tapsia

Fetens Ircina
?
hairef(C21)

Fetosa
Anthemis cotula L. Stinking Chamomile, Mayweed
camomille(C21)

Ficus [f. liquida, f. mollis]
Ficus carica L. Common Fig
figys(A3)
gallice et anglice figes(B2)
fyge(C2) anglice comyn fygge(C26)

Figula Lingue
?*Sempervivum tectorum* L. ? Houseleek
housleke(C21)

Filago
?*Hieracium pilosella* L. ? Mouse-Ear Hawkweed
?*Filago germanica* (L.) L. ? Cudweed
?*Gnaphalium uliginosum* L. ? 'Marsh Cudweed'
mouser(A2) mousere(A3)
gallice penusette, anglice weyhore(B2)
feldwort, horewort(C1) anglice horworte(C3) gallice pelusce[t]e, anglice

horwort(C6) gallice pelusete, anglice horswort(C8) gallice ganwette, anglice weyhore(C20) feldwort, horwort, schannet(C21) anglice mousere(C22) harewort [corr. horewort](C27, C28) weyhore and whyt herbe(C31) chawnette(C33) anglice feldewort, horewort or channet(C41) horewort or chaunede oþer weyhor(C42)
Lit. SB 21; Alph., p. 66, 19. See **Flores Menphis**

Filiantros
Arctium lappa L. Burdock
bur, clote(C21)
Lit. André I, p. 247 (philantropos) & II, p.196

Filipendula
Filipendula vulgaris Moench Dropwort
Filipendula ulmaria (L.) Maxim. Meadow-Sweet
filipendre(A3)
gallice filipendre, anglice stonwort(B2)
wropewort, fylypendial(C1) gallice filipender, anglice stonwort(C5) gallice filipendule, anglice stonworthe(C6) filipendule(C21) anglice rede stonebreke or rede saxifrage(C26) stonwort(C27) wild worte*(C29) stonwort(C30) cropwort [corr. dropwort](C38)
Lit. N178

Filix [filex, filix maior, filix minor]
Dryopteris filix-mas (L.) Schott Male Fern
Pteridium aquilinum (L.) Kuhn Bracken, Brake
Polypodium vulgare L. Common Polypody
gallice feuger(A1) gallice feugere, anglice fern(A6) gallice feuger, anglice brake(A8) gallice fuger, anglice braken(A9) farn(A11) gallice fugere, anglice brakes(A12)
gallice feuger, anglice fern(B1, B2) (minor) anglice polipodie(B1) polipody, everwane[sic](B4) gallice feugere, anglice fern(B5) (campestris) anglice fern (falgaria) anglice fauger(B7) gallice feuger, anglice bracan(B9)
(quercina) everfarn(C2) farn(C2) anglice bracone(C3) (campestre) anglice fern (falcaria) gallice feuger(C6) brake fern(C9) (quercina) anglice polipodie(C12) (quercina) braken, fuger de chene(C15) brake(C21) (media) osmonde(C26) anglice eververne or oke-verne or braken(C26) (minuta) overfern(C28)

Filix Arboratica/Arborea [f. arboretica]
Polypodium vulgare L. Polypody
gallice fugerole de chen(A2)

evefern(B5) feugerole de chene, vern upan oke(B8)
akeferne(C3)
Lit. see **Arboratica**

Filofolium
?*Potentilla reptans* L. ? Cinquefoil
quinfoile(C21)

Filopes
Marrubium vulgare L. White Horehound
(filolupes) horehoune, maroil, houndeberi(C21)
Lit. André I, p. 247 (philopaes) & II, 196

Fisalidos
?*Filipendula vulgaris* Moench ? Dropwort
?*Filipendula ulmaria* (L.) Maxim. ? Meadow-Sweet
snapegresse(C1)
Lit. cf. André I, p.144 (fysalis) & II, p.107; Alph., p. 66, 1; N179

Fistula
Hyoscyamus niger L. Henbane
chenillé .i. henbane(C7) gallice chenillé, anglice henebane(C12)
henbane(C24)

Fistula Bufonum
?*Filipendula vulgaris* Moench ? Dropwort
?*Filipendula ulmaria* (L.) Maxim. ? Meadow-Sweet
philipendula .i. houle-pipe(C35)

Fistula Regia
Asphodelus ramosus L. Asphodel, 'King's Spear'
Asperula odorata L. Sweet Woodruff
wodrove(C15)

Flamma [flamba, flambeon, flambe]
Iris pseudacorus L. Yellow Flag
anglice gladone(A5)
gallice glaol, anglice gladen(B1)
anglice gladene(C12)

Flammula [framula, f. aquatica, flame]
Iris pseudacorus L. Yellow Flag
Ranunculus flammula L. Lesser Spearwort

sperewort(A4) anglice sperwrt(A6) anglice sperewort(A8) anglice sperewort(A9) anglice sperwort(A12)
gallice flam, anglice gladene(B1) gladene(B2) sperwort(B3) sperworte(B4) anglice sperwort(B9)
sperwort, weyhore(C1) anglice gladene(C6) sperwort(C20) (aquatica) water-sperwort(C32)

Flammula 2
?*Mercurialis perennis* L. ? Dog's Mercury
?*Chenopodium bonus-henricus* L. ? Mercury, All-Good
anglice smerworte(C3)

Flammulus [framulus]
?
shepeswort[sic](C10) slepwort(C11, C27) sleveworth(C24) slepworte(C40)

Flaura [flauria]
?*Alchemilla vulgaris* L. ? Lady's Mantle
?*Fumaria officinalis* L. ? Fumitory
a type of clover?
medilwort [corr. mesilwort](C21) anglice fume of þe erth(C26) herba leprosum, meselwort(C32) wodrofe*(C36)
Lit. Alph., p. 63, 4; SB 21

Flores Menphie [f. menphis]
?*Hieracium pilosella* L. ? Mouse-Ear Hawkweed
filago, weyhore(C1) gallice peuusette, anglice wayhore(C16) gallice peuusette, anglice weyhore(C18)

Flos Campi
?*Silene* L. incl. *dioica* (L.) Clairv. ? '(Red) Campion'
Agrostemma githago L. Corn Cockle
(minor, ortulanus) sanegre (maior domesticus) idem(C1) (maior) sanigre (media) populе (minor) lesse sanigre (agrestis) wild sanigre, cokkil(C21) sinagria, sanagyr, a purpul floure and his sede in cappis lyke cokyl(C31) sanagre(C32)
Lit. Agnus Castus p.173 [200, 5ff. sub lolium, reading wyldsauagre]

Flos Paradisi
?
flour of paradise(C21) flowr paradise(C38)

Flos Siriacus [f. seriacus, f. ciriacus]
Althaea officinalis L. — Marsh Mallow
gallice flour de mauve(B2)
flour of malwes(C10) gallice flour de malve, anglice flour of hok(C16)
Lit. Alph., p. 67, 11; N179f.

Flosmus [flumos]
Verbascum thapsus L. — Great Mullein, Feltwort, Aaron's Rod
moleyne, veltwort(A3)
gallice moleyn, anglice feldwort(B1) gallice moleyn, anglice felwort(B2) moleyne(B3)
feltwort(C2) anglice melone [corr. molene], felwort(C3) gallice (gro) moloyne, anglice feltworth(C6) gallice moleyne, anglice feltwort vel hege-tapour(C8) molyne, sorel*(C10) gallice moleyn, anglice felwort(C16) feltwort, moleyn, lungwort(C21)

Folium
Eugenia caryophyllata Thunb. — Cloves
Dianthus caryophyllus L. — Clove Gilliflower, Clove Pink
anglice þe levys of the bladis of gelofer(C26)
Lit. Alph., p. 68, 18 'Folium quando simpliciter ponitur de folio gariofilorum intelligitur'

Folium rubi
Rubus L. ssp. — Bramble
anglice rede-brere-leves(B7)
anglice rede-bryre-leves(C6) brimbilles(C23) red-brere-leves(C28)

Fraga [fragum]
Fragaria vesca L. — Wild Strawberry
fraser(A11)
gallice frese, anglice st[r]auberie(B1) fraser(B5)
strawbury(C28)

Fragraria [fragaria, fragarium]
Fragaria vesca L. — Wild Strawberry (plant)
gallice fraser, anglice streberiwyse; gallice freses, anglice streberies(A1) freses, streberies(A3) gallice fraser(A4) gallice frese, anglice streberie(A6) gallice freses(A8) gallice fresere, anglice straberye(A9) gallice freyser, anglice strauberyes(A12)
gallice fraser, anglice strauberiewise(B2) streberiwise(B3) frasere, strawbery(B4)

streberye; streberywyse(C1) fraser, anglice straweberye-leves or straweberie-tre(C2) anglice straberywythes(C3) anglice strawbery-leves(C12) gallice frescere(C41)

Fragrasseres
Fragaria vesca L. Wild Strawberry
strauberies(B2)
strabberus[?](C1) strawberies(C16) strauberies(C18)

Fragrum [fragrun]
Ruscus aculeatus L. Butcher's Broom, Knee Holly
boneworte(C10) bonworte(C24) bonwort(C27) bruscus, anglice bonworte(C40)

Frasearia [frasdrea, frasare]
Fragaria vesca L. Wild Strawberry
ffrasere, streberiewyse(A3)
strebiri(B8)
fraser, streberywise(C22)

Frassolium
Fragaria vesca L. Wild Strawberry
strebery(C29)

Fraxinaria
?*Aegopodium podagraria* L. ? Goutweed, Ashweed
?*Cicuta virosa* L. ? Cowbane
wodewhistell(C21) ameos, skyrewhite*(C37)
Lit. Alph., p. 62, 35; B III, 259

Fraxinus [fraccinus]
Fraxinus excelsior L. Common Ash
?*Prunella vulgaris* L. ? Self-Heal
?*Ajuga reptans* L. ? Bugle
gallice freyne, anglice aysse(A1) freyne, asse(A3)
gallice freyn, anglice hasche(B1) gallice freyne, anglice asche(B2) anglice assh(B7)
(maior) hertwort, selphele(C1) asshe-tree(C2) gallice freyn, anglice asche(C6) anglice haythe-tre [read hayche-tre?](C12) hertewort, wodbroun(C21) (maior) hertwort ye more(C21) ash-tree(C29) hertwort, wodebroune(C32) asche(C35) hertwort or woddebrome(C41) hertwort oþer [w]edebrenne[sic] oþer silfehale(C42)
Lit. Agnus Castus p. 224

Fregeona
?*Phyllitis scolopendrium* (L.) Newm. ? Hart's-Tongue Fern
hertistung(C21)

Fu [feu]
Valeriana officinalis L. Valerian
valerian(A3)
gallice et anglice valerian(B2)
maturell(C3) fumetere*(C10) anglice valerione[sic](C12) lovewort, valerian(C21)

Fucus
Isatis tinctoria L. Woad
gallice wayde, anglice wode(B1)
wode(C28)
Lit. B III, 242

Fuga Demonum [f. demonis]
Hypericum perforatum L. St. John's Wort
herbe Johan(A3)
herbe Jon(B3) herb Jone(B4)
herbe John(C21, C28) anglice herb John or saint Johaniswort or wyld hoppe(C26)

Fumus Terre
Fumaria officinalis L. Fumitory, 'Earth Smoke'
gallice fumetere(A1) fumterie(A3) gallice fumetere(A4) anglice femetere(A6) fumetere(A8) fumetere(A9) gallice fumiterye(A12)
anglice fimterre(B1) gallice fumetere, anglice erþesmok(B2) fumitere(B3) fumiter(B4) fumter(B6)
(f.t. maior) fumiterre þe more (f.t. minor) lasse fumytere(C1) fumiterre(C2) anglice fumyterre(C6) gallice fimter, anglice erthesmok(C16) gallice fumeter, anglice erthsmoke(C18) (minor) lesse fymtory(C21) houndhore [corr. horehound?], fymter(C27) femter(C31)

Fundaris [fundacis]
Myrica gale L. Bog Myrtle, Sweet Gale
gallice gauge, anglice gauele(B1)
gallice gauge, anglice gaul(C6) colwort*(C28)

Fungus
var. agaricaceae Mushroom, Toadstool, Fungus
anglice tundur(B7)

cukestole vel stalestole(C3) musceroun, anglice paddokstole(C16) anglice wolfist or mescherons or taddechese(C26)

Gaisdo [gaido]
Isatis tinctoria L. — Woad
Reseda luteola L. — Dyer's Rocket, Weld
wolde(B5) wode(B5)
waulde(C14) welde(C37)

Galanga [galangia]
Cyperus longus L. — Galingale, 'English Galingale', 'Sweet Cyperus'
?*Alpinia officinarum* Hance — ? Galangal, China Root
?*Alpinia galanga* (L.) Willd.
gallice galingal(A1) galingal(A3)
gallice et anglice galingale(B1) gallice et anglice galingale(B2) galyngale(B3)
(g. maior) more galyngale (g. minor) lasse galingale(C1) anglice galyngale(C6)
Lit. N181

Galaustiana
Gentiana amarella (L.) Börner — Felwort
gentiana, gencyan, baldemony, karyswhete(C31)

Galbanum
Ferula galbaniflua Boiss. et Buhse — Galbanum
þe juse of a herbe(C21)

Galla
Quercus L. — Oak Apple, Oak Gall
gallice pome de cheyne(A1) poume de cheyne, hokappel(A3) ocappel(A7) anglice akapyl(A10)
anglice okappil(B1) gallice poume de cheyne, anglice apel of þe hok(B2) okappel(B3)
oke-appel(C10) gallice pome de cheyne, anglice okeappyll(C16) pomme de kene, oke appull(C22) hok-happel(C35)

Gallellus
Teucrium scorodonia L. — 'Sage-Leaved Germander', Wood Sage
ambrose, wild sawge, hyndeheyle(C21)

Galli Cresta
Salvia sclarea L./*horminoides* Pourr. Clary/Wild Clary
clareie(B3)

Galli Iratos [for gallitricus?]
?*Eupatorium cannabinum* L. ? Hemp Agrimony
?*Ranunculus aquatilis* L. ? 'Water Liverwort'
waterwort(C21)
Lit. B III, 248

Gallia Diaria [for gallia dana]
Bryonia dioica Jacq. Bryony
wild nepe(C21)
Lit. André I, p.146 (galiadana)

Gallitricos [gallitricis]
Adiantum capillus-veneris L. Maidenhair-Fern
maidhere(C21)

Gallitricum [gallitrum]
Salvia sclarea L./*horminoides* Pourr. Clary/Wild Clary
Agrostemma githago L./*Lolium temulentum* L. Cockle/Darnel
Capsella bursa-pastoris (L.) Medic. Shepherd's Purse
bledewort vel yetewort(B1) anglice godes-eie(B2) popil, nele(B4) (g. maius) galli crista, clareye(C1) crista galli, popill(C7) anglice schalwort(C8) anglice cristis-eye(C16) shepperdis-purs(C21) blodewort(C28) colkis-camb(C35) gallitrik(C37)

Gallitricus [gallitricum]
?*Eupatorium cannabinum* L. ? Hemp Agrimony
?*Asplenium trichomanes* L. ? Maidenhair Spleenwort
?*Ranunculus aquatilis* L. ? 'Water Liverwort'
feterwert(A11)
waterwort(B3)
Lit. B III, 248 see **Galli Iratos**

Garantia
Rubia tinctorum L. Madder
warente, madre(C21)

Gariofilata [gariofilatum]
Geum urbanum L. — Avens, 'Clove Root (Radix caryophyllata)'
gallice avans(A1) avance, harefot(A3)
gallice avence, anglice harefot(B2) avence(B3) gylfoyle(B4) clouwus(B5) avance, anglice haselwort*(C2) anglice avence vel harefot(C8) anglice clowes(C12) (agrestis) avance sauv[a]ge(C12) gallice avance, anglice horsfote [corr. harefote](C16)

Gariofilus [gariophili]
Eugenia caryophyllata Thunb. — Cloves
Dianthus caryophyllus L. — Clove Gilliflower, Clove Pink
?*Geum urbanum* L. — ? Avens
gallice gilofre(A1) gilofre(A3) gallice gilofre(A4)
jelofre(B1) giloffre(B3)
clowe(C1) gilofre, clouwe(C2) anglice gylofre(C6) anglice clowe, gilofre(C12) klowes(C29)

Garsaria
?*Lolium temulentum* L. — ? Darnel
anglice darnale(C29)

Gazarus [gazara]
?*Polygonum aviculare* L. — ? Knotgrass
?*Centaurea scabiosa* L. — ? Greater Knapweed
þe grete knotwort, gazer(C9)

Gelena
Citrullus colocynthis (L.) Schrad. — Colocynth
Lagenaria vulgaris Ser. — Gourd
wild nepe(C21)
Lit. André I, p.146 (gelala) & II, p.109; Alph., p. 73, 16; N182

Genesta [genusta, genustula, genestula, genestia, genestum]
Sarothamnus scoparius (L.) Wimmer ex Koch — Broom
gallice geneste, anglice brome(A1) geneste, brom(A3) anglice brom(A8)
gallice geneste, anglice brom(B1) gallice genise[sic], anglice brom(B2) chamgk(B3) anglice brom(B3) genette, benet, brom(B8) (g. maior) [b]rome (g. minor, genestula) lasse brome(C1) brom(C2) brom, genestre(C21) is liche a brome with a whyte flour(C27) anglice genesture or browneworte(C41)

Genicularis
?Polygonum bistorta L. ? 'Bistort'
Hyoscyamus niger L. Henbane
(maior) þe gre[te] genycle, þe grete knewort, þe mene violet (see in bisaco)(C9) (minor) þe litle bistort, the litle genycle(C9) anglice hennebane(C22)
Lit. cf. André I, p.147 & II. p.109

Geniculata
Polygonum aviculare L. Knotgrass, 'Swine's Grass'
swynesgras(B3)
swynesgrece(C7)
Lit. cf. André I, p.147 & II, p.109

Genocelurum
?Anchusa arvensis (L.) Bieb. ? Bugloss
lang de bef(C21)

Gentiana
Gentiana amarella (L.) Börner Felwort
gallice genciane, anglice carswete(A1) genciane(A3) gallice genciane(A4) veldwert(A11)
genciane, anglice feltewort(B1) gallice genciene, anglice baldemoigne(B2) genciane, feldwort(B3)
genciane, anglice baldemoyne(C1) baldemoyne(C2) anglice gentiane vel crasuet(C5) genciane, carswete(C7) anglice gentian vel crasuet(C8) karswete(C14) gallice gencian, anglice baldmony(C16) baldmony(C21) genciani[sic], carsuet(C23) felewort(C28) bishopworte*(C29)

Gerateon
?Lactuca virosa L./*serriola* L. ? Wild Lettuce/'Prickly Lettuce'
wild letuse(C21)

Germandrea
Teucrium chamaedrys L. Common or Wall Germander
?Cymbalaria muralis Gaertn., Mey. & Scherb. ? Pennywort, Ivy-Leaved Toadflax
gallice germandre(A1) germandre(A3)
gallice germandre, anglice medratele(B2) peniwort(B3) germandir(B3)
germandre(C2) anglice mederatill(C16) (maior) penygres(C21) medeueratil(C23) metratil(C36)

Gernicella
?*Sedum acre* L. ? Stonecrop
stoncrop(B3)

Gersa
?*Agrostemma githago* L. Corn Cockle
?*Lolium temulentum* L. Darnel
kokkull(C24)
Lit. cf. Alph., p. 74, 14

Gesamselos [for ges omfalos]
Cymbalaria muralis Gaertn., Mey. & Scherb. Ivy-Leaved Toadflax, Pennywort
?*Umbilicus rupestris* (Salisb.) Dandy Pennywort
penywort(C21)
Lit. André I, p.148 & II, p.110

Giddea
Anthriscus cerefolium (L.) Hoffm. Chervil
anglice cerefoyle(C12)

Gira Solis
Leonurus cardiaca L. Motherwort
moderwort(B3)
moderwort(C2) anglice moderworte(C12)

Giros
?
peletre(C21)

Girum Solis [gramen solis]
Lithospermum officinale L. Gromwell
gromil(C1)

Git [gith]
Agrostemma githago L. Corn Cockle
Lolium temulentum L. Darnel
gallice neel, anglice cokel(A1) kokkele, neel(A3)
gallice nel, anglice cockel(B1) anglice cockel(B2) cokkel(B3) cokell(B6)
nele, cokkell, popill(C7) anglice kochull(C16)

Glacias [for glaucia?]
Chelidonium majus L. Greater Celandine

?*Glaucium flavum* Crantz/ *corniculatum* (L.) Rudolph ? 'Yellow/Red Horned-Poppy'
teterwort, celidony(C21)
Lit. Alph., p. 36, 5 ('glaucia') See **Glaucium**

Gladiolus [gladiolum]
Iris pseudacorus L. Yellow Flag, Gladden
Acorus calamus L. Sweet Sedge
seg(A3) gallice flaume, anglice gladene(A5)
(gladiolus plaustris) .i. leverwort(B1) gallice gladene, anglice segge(B2) gladene(B3)
gladiole, segge(C1) anglice gladyn(C3) gallice glayole, anglice gladene vel segges(C6) suchegresse(C24) gladyn vel segges(C28) gladewyne(C40) anglice gladioll or gladew(C41)
Lit. cf. B III, 157 (l[to]fer)

Glaucium [glaucia]
Chelidonium majus L. Greater Celandine
?*Glaucium flavum* Crantz/ *corniculatum* (L.) Rudolph ? 'Yellow/Red Horned-Poppy'
anglice celydoyne(C10) anglice celidoyne(C12) (agreste) wild celidony(C21)
Lit. Alph., p. 36, 5; André I, p.150 (glaucium) & II, pp. 111f.

Glaucus
Crocus sativus L. Saffron, Crocus
Carthamus tinctorius L. Safflower, Bastard Saffron
saferoun(C24)

Glechon [gliconicum]
Mentha pulegium L. Pennyroyal
Thymus serpyllum L. Wild Thyme
puliol(A3)
pulyole(C22)
Lit. André I, p.150 (glechon) & II, p.112. See **Gliconium**

Gletinaria
Arctium lappa L. Burdock
clote(C23)

Gliceria [glicoricia, liquericia, liquiricia, glicea, gliscia, gliciata]
Glycyrrhiza glabra L. Liquorice

gallice licorise(A1) licorise(A3)
gallice et anglice licoris(B2) licoriz(B3) licorice(B6)

Glicida [glicia]
Paeonia mascula (L.) Mill. Peony
pioyne(B3)
anglice pyony(C3) pyony(C4) anglice pyonye(C26)
Lit. André I, p.151 (glycis/glycyside) & II, p.112; Alph., p. 76, 16

Gliconium
Mentha pulegium L. Pennyroyal
Thymus serpyllum L. Wild Thyme
puliol reale(C10) polygonia, a clote*(C19) pulioll real or churchewort(C21) anglice þe holle-ryd or majeron or churchewort(C26)

Glis
Arctium lappa L. Burdock
gallice gletuner, anglice clote(A1) glettener, clote(A3)
anglice scallegres(B2) borre(B3) clote(B5)
scalgras(C16) scalegrasse(C18) scalle[grasse](C23)

Glossera
?*Smyrnium olusatrum* L. ? Alexanders, Horse-Parsley
alisaundyr(B8)

Glutum
Arctium lappa L. Burdock
romani vocant glutumere [read glutinnere?], anglici clote(C43)

Gnicus [gernicus, gliccus]
Carthamus tinctorius L. Bastard Saffron
Crocus sativus L. Saffron
gallice saffrane(A1) safran(A3)
gallice et anglice saffran de curtil(B2)
gallice et anglice safrun d'orte(C16) gallice et anglice saufron de curtyl(C18)
Lit. André I, p. 94 (cnecos) & II, p. 69; Alph., p. 75, 25

Golata
?*Avena fatua* L. ? Wild Oat
wylde ote(C23)

Golena
Quercus L. Oak-Apple
anglice oc-appel(C12)

Graciosa
?*Linaria vulgaris* L. ? Toadflax
?*Gratiola officinalis* L. ? Hedge Hyssop
graciose(B3)

Gramen
Agropyron repens (L.) Beauv. Couch-Grass, Scutch, Twitch
anglice quiche(A1)
anglice quiche(B2) anglice quichene(B7)
grasse(C1) gythe*(C3) gallice et anglice [sic] quich(C16) anglice grete grasse or qwich-penne(C26) gras vel quichen(C36)
Lit. B III, 66f.

Gramen Amesistis [granis a.]
Agropyron repens (L.) Beauv. Couch-Grass
anglice cuchegras(C10) chuchgras(C27) cuchegras(C40)

Granum Diureticum [gramen d.]
Lithospermum officinale L. Gromwell
anglice hyneale [corr. lychewale](A6) anglice lyewal(A8) anglice liȝewal(A9) anglice lyþewal(A12)
lychewal(B3) anglice licchewal(B5) anglice lythewale(B9)
lychewale(C1) anglice lyghtwall(C3) anglice lighwale(C3) anglice lichewale(C12) lythewort(C29) lichewall(C33) lyghwhall(C34)
Lit. Alph., p. 72, 27

Granum Paradisi
Lithospermum officinale L.
Gromwell wylde gromyll(C41)

Granum Solis [grana s.]
Lithospermum officinale L. Gromwell
gallice gromyle(A1) gromoil(A3) anglice gromyle(A4) anglice gromylle(A6) gallice grommil(A8) gallice gromoil(A9) gallice gromoyl(A12)
gallice et anglice gromil(B2) gromyl(B3) gallice gromyl(B5) grumyl gallice(B5) sossie*(B5) grumyle(B6) gallice et anglice wild gromil(B9)
(maius) gromel (agreste, minus) wyld gromel(C1) (agrestis) wild gromyll(C21) gromole(C24) (agreste) wilde gromel(C32)

goldwort*(C35) anglice gromyll or lytyll wale [corr. lichwale](C41)
Lit. Agnus Castus p. 225 (little wale)

Grassa
Mercurialis perennis L. Dog's Mercury
Chenopodium bonus-henricus L. Mercury, All-Good
mercurie, papwort or more smerwort(C21)

Gratia Dei
?*Geranium sanguineum* L. ? 'Bloody Cranesbill'
?*Calendula officinalis* L. ? Marigold
Gratiola officinalis L. 'Hedge Hyssop'
seynt Marie rode(B1)

(maior) more grace dieu (minor) lasse grace dieu(C1) (minor) muser*(C35) grace of god þe more(C38) gratia dei þe more and þe lasse(C41)
Lit. cf. Alph., p.72,15; SB23. See **Deo Gratias**

Grias
Rubia tinctorum L. Madder
mader(B3)
Lit. B III, 167 & 260; André I, p.153 & II, p.113

Grosiler [grosilere, gresiler]
Rhamnus catharticus L. Buckthorn
?*Ribes grossularia* L. ? Gooseberry
anglice thewthorn(B7)
anglice thevethorne(C5) thethorn(C7) þeveþorn(C23) thethorn(C36)
Lit. B III, 236; L1268

Gruncus [grunca]
Cuscuta epilinum Weihe Flax Dodder
anglice doder(C12) dodre(C21)
Lit. Alph., p. 77, 27

Hasta Regia
Asperula odorata L. Sweet Woodruff
Asphodelus ramosus L. Asphodel
anglice woderove(A2) anglice woderove(A6) woderove(A7) woderouve(A8) anglice woderowe(A9, A12) anglice woderove(A10) woderove(B3) wderove(B5)

anglice woderoufe(C3) woderove(C19) woderof(C21)
Lit. B I, 146f & III, 267f

Hastula Regia [astaragia, agula r.]
?*Hypericum perforatum* L. ? St. John's Wort
Asperula odorata L. Sweet Woodruff
herbe Johan(A3)
gallice muge terrestre, anglice wodrofe(B1) gallice muge dé boys, anglice woderove(B1) anglice woderove(B2) gallice muge dé bois, anglice woderove(B2) wodrowe(B4)
woderove(C1) shawewort(C21) muske bastarde, woderove(C37)
Lit. André I, p.157 & II, p.117

Hastula Sancti Christofori
Actaea spicata L. Baneberry, Herb Christopher
scabwort(C10) pollicaria minor, scabbeworte(C24) schawewort(C27)

Hedera Arborea
Hedera helix L. Ivy
ere, yvy-tre(A3) gallice ere, anglice yvy(A6) yvi(A8)
gallice er, yve[sic], anglice yvi(B2)
yvy of treys(C1) tre-ywyn(C3) gallice er, ive, anglice yvi(C18) anglice wodebend-yvy(C20) ere, tre-yvy(C22)

Hedera Muralis
Hedera helix L. Ivy
yvy of wall(C1) wall-yve(C38)

Hedera Silvestris
Hedera helix L. Ivy
anglice yvi(A2)
yvy(B3)
wode-ivy(C28) wod-ivy(C41)

Hedera Terrestris
Glechoma hederacea L. Ground Ivy
anglice heyove(A2) herbe [corr. erre] terestre, heyhove(A3) heihove(A8) anglice hove(A9) anglice ayove(A10)
ether-yvi(B1) anglice grund-hivi vel wodewynd(B1) gallice ere terestre, anglice heyhove(B2) heyhove(B3, B5) grond-ivy(B8) anglice tounhove(B9)
(alba) wyt orþe-yve (nigra maior) þe more hayhove (minor) lasse hayhove(C1) erthe-yvy(C2) anglice erdh-ywyn(C3) anglice[sic] grant

eyre vel heyrith* vel heyhove(C8) hayhove, tunhove(C24) grounde-ivy(C28) grund-ivi(C35) anglice ȝerdhove vel heyhove vel alhowe(C36) (alba) whitt erþe-yve(C38) anglice erþ-yve, heyhove or erth-ivy(C41)

Heliotropia [elitropia, elentropia, elitropium, eliotropus]
Calendula officinalis L. Marigold
anglice rodflour(C8) wrotwort(C9) solsekil, roden(C10) marygoolde(C21) solsequile(C24) gooldis(C28)
Lit. cf. SB 43 ('verucaria .i. solsequium': = *Caltha palustris* L. 'Marsh Marigold'); B III, 212; André I, pp.158f & II, p.119

Heptafilos [eptafilum, eptafilon]
?*Potentilla erecta* (L.) Räusch. ? Common Tormentil
?*Scrophularia aquatica* L. ? Water Betony
brunwrt(A11)
elmewort, loudwort(C28) sevebefe [corr. sevelefe](C28) seaveleaf(C37)
Lit. cf. B I, 24 and III, 41; André II, p.120; B III, 205 (seafonleafe)

Herba Acetosa
Rumex acetosa L./*acetosella* L. Common Sorrel/Sheep's Sorrel
Oxalis acetosella L. Wood Sorrel, Wood Sour, Stubwort
anglice surdoke(A2) anglice sourdok(A6)
sourdokke(B3) anglice surdocke(B5) surdock(B5) anglice surell(B7) anglice souredok(B9)
souredok(C7) sorel or souredok(C10) cokesowre, sorel de boyse, stobwort(C21)anglice sorel dé boys(C36)

Herba Angelica
Lamium album L. White Dead-Nettle, Archangel
dede-nettill, archaungel(C21)

Herba Benedicta
Conium maculatum L. Hemlock, Herb Bennet
herbe benet, hemeloc(A3)
gallice herbe beneit, anglice homeloc(B2) hemelok(B3) herbe benet, hemelok(B4) hemelok(B5)
hemelok(C1)

Herba Campana
?*Solanum nigrum* L. ? 'Black Nightshade'
morel(C2) anglice morel(C12)

Herba Cancri
?
camomylle(C21)
Lit. cf. Alph., p. 38, 18

Herba Canicularis – see **Canicularis**

Herba Cassica
?*Mercurialis perennis* L. ? Dog's Mercury
?*Chenopodium bonus-henricus* L. ? Mercury, All-Good
anglice mercurie(C12)

Herba Catholica
Laurus nobilis L. Laurel
laureole(C12) hemelok*(C21) laureol(C21) anglice laureal-sede(C26)
Lit. Alph., p. 79, 11

Herba Certa
Achillea millefolium L. Yarrow, Milfoil
anglice yarow or noseblede or mylfoyl(C26)

Herba Christophori [h. sancti c.]
?*Actaea spicata* L. ? Baneberry, Herb Christopher
?*Pulicaria dysenterica* (L.) Bernh. ? Fleabane
?*Pulicaria vulgaris* Gaertn. ? 'Small Fleabane'
?*Lysimachia vulgaris* L./ *thyrsiflora* L. 'Yellow/Tufted Loosestrife'
?*Osmunda regalis* L. ? Royal Fern
scabgras(A3) anglice scabgrasse(A5)
scabewurt(B1) gallice herbe Cristofore, anglice scabbegres(B2) herbe Cristofer(B4)
anglice scabwort(C5) anglice schabbegras(C6) gallice[sic] myldemad, anglice scabgres(C8) scabwort(C10) anglice schalgres(C16) gallice herbe Cristofore, anglice scabgrasse(C18) herbe Crystofore(C31) schallegrass(C36) herbe Cristofre male, herbe Cristofore femele(C41)

Herba Cotidon
Cymbalaria muralis Gaertn., Mey. & Scherb. Ivy-Leaved Toadflax
penywort(B3)
anglice peniwort(C12)

Herba Cruciata
?
anglice sporge*(C28) spergula maior, ippia minor, chicunmete; spergula minor, due sunt species: croswort, herba Pauli(C32) spergula maior, anglice croyser(C36) anglice croysewort or exan(C41)
Lit. Agnus Castus, pp. 247f; p. 227 (exan)

Herba Dentaria
Anacyclus pyrethrum DC. — Pellitory
Parietaria diffusa Mert. & Koch — Pellitory-of-the-Wall
peletre(B4)
anglice peletre(C12)
Lit. André I, p.117 (dentaria) & II, p. 88

Herba Deorum
Teucrium scorodonia L. — Wood Sage, Ambrose, Hindheal
wild sawge, ambrose, hindeheile(C21)

Herba Diana
Artemisia vulgaris L. — Mugwort
mogwort(B3)
anglice mogwort(C12)
Lit. André I, p.118 (Dianaria herba) & II, p. 88

Herba Fetida
?*Aquilegia vulgaris* L. — ? Columbine, Culverwort
dowesfote(C21)

Herba Fullonum
Saponaria officinalis L. — Soapwort
Dipsacus fullonum L. — Teasel
anglice crosehope(C16) crowsope or tasell(C21)
Lit. N183

Herba Herçulis
Sempervivum tectorum L. — Houseleek, Sengreen
?*Sedum* L. ssp. — ? Stonecrop
anglice syngrene(C8)
Lit. André I, p.160 (herba hirculi) & II, p.121

Herba Hircina
?*Galeopsis tetrahit* L. — ? 'Common Hemp-Nettle'

anglice swannestong(C41)
Lit. Alph., p. 80, 2

Herba Ivonis
?*Plantago coronopus* L. — Buck's-Horn Plantain
?*Ajuga chamaepitys* (L.) Schreb. — Ground-Pine
herbyve(B3)

Herba Judaica
?*Tragopogon pratensis* L. — ? Goat's-Beard
?*Galium aparine* L. — ? Cleavers, Hayrif
hairef(C21) herba ircina, -fetens, anglice wyld wacches or wylde mospeyse(C26)
Lit. cf. Alph., p. 83, 17

Herba Leprosorum
?*Alchemilla vulgaris* L. — ? Lady's Mantle
?*Fumaria officinalis* L. — ? Fumitory
medelwort [corr. meselwort](C21) planta leonis, flaura, meselwort(C32) meselwort(C38)

Herba Linaria [h. lunaria; for luninaria?]
Linaria vulgaris L. — Toadflax
Verbascum thapsus L. — Great Mullein, Aaron's Rod
gallice linere(A5)
molyne(C10) moleyn(C21)

Herba Luminaria [h. luminaris]
Verbascum thapsus L. — Great Mullein, Aaron's Rod
moleyne(B3)
moleyn, feltwort(C7) anglice moleyne(C12) cattstayle or feldewortte or soft(C41)
Lit. Macer pp. 222f.

Herba Lune
?*Botrychium lunaria* (L.) Sw. — ? Moonwort
?*Aster amellus* L. — ? Aster
lunari(C21)

Herba Martis [h. mortis]
?*Lilium martagon* L. — ? 'Martagon Lily'
?*Portulaca oleracea* L. — ? Purslane

mortagon(C21) (h. mortis maior, martis masculina) mortagon (h. mortis minor, martis feminea) lasse mortagon(C38)

Herba Mede [h. medee]
?*Melissa officinalis* L. ? Balm
medewort(C10, C21, C27) medeworte(C24)

Herba Memoria [MS correction from Marmoria]
Artemisia absinthium L. Wormwood
wormode(C21)

Herba Militaris
Orchis militaris L. Soldier Orchid
?*Endymion non-scriptus* (L.) Garcke ? Bluebell, Wild Hyacinth
anglice wolvyskul or wolvysballock or harebelles(C26)

Herba Munda
Euphorbia L. spp. Spurge
anglice spurge(A2) spurge(A3)
sporge, g(l)utcorn(C8)
Lit. B III, 112f.

Herba Muralis [h. mirabilis]
Parietaria diffusa Mert. & Koch Pellitory-of-the-Wall
porrettorye, litillwort, hemerwort(C21) gallice culray[?](C37)
Lit. Alph., p. 80, 8

Herba Muscata
Asperula odorata L. Sweet Woodruff
anglice woderef(A4)
muge dé boys(C5)
gallice myld[sic] dé boys, anglice wodrof(C16)

Herba Paralisis [h. paralitica, h. paralisie]
Primula veris L. Cowslip
Primula elatior (L.) Hill Oxlip
Primula veris L. x *vulgaris* Huds. Common Oxlip
anglice cuslope(A8)
cusloppe(B1) gallice primerole, anglice couslop(B2) anglice cusloppe(B5)
oxlippe(C7)

Herba Pauli
?*Spergula* L. ssp. ? Spurrey
herbe Poule(C21) spergula minor, herbe Poule(C32)

Herba Pedicularis [h. pedicialis]
Delphinium staphisagria L. Stavesacre
staphisagre(C21) stavsauge[sic](C28)
Lit. André I, p. 241 (pedicularia herba) & II, p.190; N184

Herba Perforata
Hypericum perforatum L. St. John's Wort
gallice herbe s. John, anglice s. Johannis erbe vel s. J. wort(C16)
Lit. N184

Herba Petri
Primula veris L. Cowslip
herbe Petre or cowslop(C21) herbe Peter .i. kowslap[sic](C32)
koweslippe, gallice herbe Peter, herba sancti Pauli, spergula(C41)

Herba Petronille
?
herbe pernell(C21)

Herba Pigmentaria
Melissa officinalis L. Balm
medewort(B3)
medsote(C2) anglice medsote(C12)

Herba Pilicum [h. publicum, pulicum, pulteum]
?*Iris pseudacorus* L. ? Yellow Flag
?*Acorus calamus* L. ? Sweet Flag, Sweet Sedge
flagges(A3)
anglice flaggus(B7)
anglice flagges(C6, C8, C22, C23, C36)

Herba Pratalis
Lonicera periclymenum L./ *caprifolium* L. Honeysuckle
Melilotus officinalis L. Yellow Melilot
Trifolium L. ssp. Clover
anglice honisucul(B7)
anglice honysoke(C6) a þre-levyd gras, cockhedis(C7)

Herba Pulicaris
Plantago indica L. — ('Plantago Psyllium') Flea Seed, Psyllium Plantain
prosill [for psill?](C21)
Lit. Alph., p. 83, 13

Herba Putica [for putidra]
Anthemis cotula L. — Stinking Chamomile, Mayweed
maithes(C14)
Lit. B III, 165f.

Herba Regia
Asperula odorata L. — Sweet Woodruff
muge savage gallice, anglice woderove(A2)
woderofe(C28)

Herba Roberti
Geranium robertianum L. — Herb Robert
?*Geranium molle* L. — ? Dove's-Foot Cranesbill
gallice roal(A2) herbe Robert(A3)
gallice et anglice herbe Robert(B2) herbe Robert(B4) herbe Roberd(B8) anglice herbe Robard(C8) erbe Robert, anglice wodewort(C16) gallice et anglice herbe Robart(C18) herbe Robret[sic] .i. colvirfote(C31) romani vocant eam herb Roberd, anglici wedewort(C43)

Herba Rotunda
?*Betonica officinalis* L. — ? Betony
anglice betonye(C12)

Herba Salamon [h. Solamen]
?
herbe solamen, ranclwort(C34) herbe Salamon, rancleworte(C37)

Herba Salutaris
Solanum nigrum L. — 'Black Nightshade'
nytischode, ilhaw, litil morell(C21)

Herba Sancte Marie
Chrysanthemum vulgare (L.) Bernh. — Tansy
Calendula officinalis L. — Marigold
tanesei(A2)
rodewort, holygold, seynt Mary gold(C7)

Herba Sancte Marie Minor
Polygonum persicaria L. — Red Shank, 'Persicaria'
Polygonum hydropiper L. — Water-Pepper, Culrage
anglice culrage(C26)

Herba Sancti Johannis
Hypericum perforatum L. — St. John's Wort
gallice herbe seynt Jon, anglice seynt Joneswort(B2) herbe Jon(B3)
anglice et gallice milepertus(B9)
alrewort(C7) hundryd-hoole(C31)
Lit. N184

Herba Sancti Laurencii
Sanicula europaea L. — Sanicle
anglice sanicle(C26)

Herba Sancti Pauli
?*Euphorbia* L. ssp. — ? Spurge
?*Spergula* L. ssp. — ? Spurrey
spurge(B4)
spurge(C7, C12) herbe poule(C27) rede dok*(C28)
Lit. Alph., p. 79, 4

Herba Sancti Petri [herba Petri]
Primula veris L. — Cowslip
Primula elatior (L.) Hill — Oxlip
Primula veris L. x *vulgaris* Huds. — Common Oxlip
anglice kuslop(A2) couslopp(A3) anglice kuslop(A5) anglice cousloppe(A6)
cousleppe(B3) premeroll(B4)
anglice primerose vel cowsleppe(C6) anglice primerose or oxlyppe or cowslyppe(C26) herbe Petre, cowslop(C27)

Herba Scelerata
Ranunculus sceleratus L. — 'Celery-Leaved Crowfoot'
?*Anthriscus sylvestris* L. — Wild Chervil
crowfote, or cerfoile(C21)

Herba Scripta
Phyllitis scolopendrium (L.) Newm. — Hart's-Tongue Fern
anglice hertystonge(C26)
Lit. Alph., p. 80, 4

Herba Silvana	
Asperula odorata L.	Sweet Woodruff
woderove(B5)	
Herba Siriaca	
Althaea L. ssp.	Mallow
Malva L. ssp.	Mallow
anglice mallowes(C26)	
Lit. see **Siriaca**	
Herba Sulphurata	
?*Peucedanum officinale* L.	? Hog's Fennel, Sulphur-Weed
horstoth(C21)	
Herba Tinctorum	
?*Galium cruciata* (L.) Scop.	? Crosswort
?*Rubia tinctorum* L.	? Madder
anglice sperge(C12)	
Herba Tribulosa	
Ulex europaeus L.	Gorse, Furze
gersk or ffirces(C21)	
Herba Trutannorum	
Thapsia garganica L.	Thapsia
faitorswede, ffatorswede(C21)	
Lit. Alph., p. 83, 16; see **Truta**	
Herba Ursina	
?*Lactuca serriola* L./*virosa* L.	'Prickly Lettuce', 'Wild Lettuce'
horsethistil(B1)	
horsþistell(C28)	
Herba Veneris	
Verbena officinalis L.	Vervain
verveyne(C21, C26)	
Herba Venti	
Centaurea scabiosa L.	'Greater Knapweed'
Centaurea jacea L./*nigra* L.	'Brown-Rayed Knapweed', 'Lesser Knapweed'
madefelon(B3)	

mathefelon, sed quidam dicunt quod est consolida media(C2) matefelon, codwede(C7) matefeloun(C19) (maior) matefeloun(C19) (maior) anglice matfelon or knapwort(C26) anglice herbe ive(C26)

Herba Veris
Primula veris L. — Cowslip
primerole(C36)

Herba Victorialis
Vinca maior L./*minor* L. — Great Periwinkle/Lesser Periwinkle
anglice pervynke(C26)

Herba Viperina see **Viperina**

Herba (Sancti) Walteri
Asperula odorata L. — Woodruff
?*Potentilla anserina* L. — ? Silverweed
gallice herbe Water(B2) smerwort(B4)
angice herbe Water(C6) Water herbe(C23) romani vocant eam herbe Water vel muge dé boys, anglici smerewort(C43)

Herba Willelmi
?*Apium graveolens* L. — ? Wild Celery, Smallage
?*Artemisia vulgaris* L. — ? Mugwort
marche(C21) mugwort(C21)
Lit. C9 = acalifa, 'nettle', C27 = amariscus

Herba Yve
?*Plantago coronopus* L. — ? Buck's-Horn Plantain
hertys-horne(C31)
Lit. Alph., p. 80, 2; p. 46, 14

Hermodactila [ermodactula, hermodactili, hermodactilus, hermodactiles, hermodactulis]
Allium ursinum L. — Ramsons, Wild Garlic
Allium vineale L. — Crow Garlic
Colchicum autumnale L. — Meadow Saffron
crouelek(B1) croulek(B3)
hermodactull(C1) crowleak(C2) crowlyke(C7) gallice hermodactil(C8) anglice crowegarlek(C12) anglice hermodactili(C12) anglice croweleik(C12) anglice molde*, betes*(C16) anglice melde*, betes*(C18) crouleke(C28) ramese(C28) romani vocant eam hermodie,

anglici galluc vel crowlek(C43)
Lit. B III, 61f & 102

Hermolia [ermola]
Peganum harmala L. Wild Rue
?*Thalictrum flavum* L. 'Common Meadow Rue'
anglice rewe(C16) anglice rwe(C26)
Lit. Alph., p. 58, 29 (erimola); André I, p. 41 (armala) & II, p. 25

Hierobotane [ierobotonum, ierobotanum]
Verbena officinalis L. Vervain
anglice et gallice verveyn(B1)
vervayne(C3) iyenharde(C28) peristerion, anglice hisene3erde(C43)
Lit. André I, p.162 (hierobotane) & II, p.123

Hinnula
Inula helenium L. Elecampane
Allium cepa L. Spring Onion
sperwort(B3)
anglice chibolles(C4)
Lit. Alph., p. 86, 10

Hipoceta
Equisetum L. ssp. Horsetail
folestaile(C21)
Lit. Alph., p. 82, 14 (hircina: hyposita . . .)

Hiposelinum [yposilina, iposilea, ipophilon, ypofilvia, ypofilia, yposilia]
Veronica beccabunga L. Brooklime
gallice faverole, anglice lemek(A1) faverole, limik(A3) (aquatica)
faverole(B1) lemeke(B3)
faverele, lemke(C4) anglice brokelempe(C2, C12) brokelemke(C19)
brokelenyk(C21) faverole, lemyk(C21) waterlilie(C35)
Lit. André I, p.163 (hipposelinum) & II, p.124; Alph., pp.197 & 86, 16

Hirema [hirena]
Equisetum L. ssp. Horsetail
folestaile or paddokpipe(C21) a row herbe with wych men pulschyn
combes, bowes and coppes(C27)
Lit. cf. Alph., p. 37, 4 (cauda pulli) & p. 82, 14 (hircina)

Hirtago
?
meald(C21)

Humilus
Sambucus ebulus L. Danewort, Dwarf Elder
walwort(C21)
Lit. cf. André I, p. 280 (sambucus parva/tenera) & II, p. 225

Hypericon
Hypericum perforatum L. St. John's Wort
herbe Johan(A3) seint Johanhisgras vel-wort(A3)
herbe Jon(B3) herbe Jone(B6)
anglice herbe Jone(C3)

Hypnotica [hermotica]
?*Chrysanthemum vulgare* (L.) Bernh. ? Tansy
tanesei(A2)
Lit. cf. Alph., p. 87, 13 (innoticum) & André II, p.128

Hysopus [isopus]
Hyssopus officinalis L. Hyssop
gallice et anglice ysope(A2, A12) ysope(A3) anglice et gallice ysap(A6)
anglice isope(A9)
ysope(B1) gallice et anglice ysope(B2) isope(B3) ysope, thringes(B4)
(h. domesticus) ysope (h. maior agrestis) wylde ysope (minor agrestis)
lasse wylde ysope(C1) (ortulanus) anglice gardyn isope(C26) (agrestis)
isope of the wood(C26)

Hyspyros
Equisetum L. ssp. Horsetail
coltystayle(C38)
Lit. André I, p.163 (hippuris)

Iacea
Centaurea scabiosa L. 'Greater Knapweed'
Centaurea jacea L./*nigra* L. 'Brown-Rayed Knapweed'/'Lesser Knapweed'
matefelun(A2) mathefoloun(A3). i. alba et nigra, anglice
madefeloun(A6) gallice matafelun(A8) alba vel nigra, anglice
matefelun(A9) anglice matfelon(A12)

madfeloun, anglice snapwort [corr. knapwort?](B2) (nigra) maudefelon(B3) matfeloun(B4) anglice madefelun(B5) anglice maidefelon(B5) matfeloun(B6)
anglice matefelon(C3) (nigra) anglice madefelon, codwede [corr. cowede?](C6) (alba) scabiose(C7) anglice snapwort [corr. knapwort?](C16) (nigra) gallice madfelon, anglice snapwort [corr. knapwort?](C18) scabiouse(C28) (maior, nigra) hard knopwede vel matyfylon (minor, alba) lasse matyfylon(C38)

Iacea Alba
Centaurea scabiosa L. 'Greater Knapweed'
?*Potentilla anserina* L. ? Silverweed, 'Goose Tansy'
matefeloun wit a purpur flowr(C10) gallice matefelon, anglice clodwort(C20) gosegresse, sikilwede, wild tansy(C21) matfelon rede(C24) matfeloun with a purpyl flour(C27) hertyshorne* and ys a spyce of scabyose(C31) gosegras, wylde tansey(C32) goose-grasse or wylde tansye(C41)

Iacea Nigra
Centaurea nigra L. 'Lesser Knapweed'
anglice matefelune(A5)
gallice madfeloun, anglice snapwort [corr. knapwort?](B2)
matfeloun wit a reed flour(C10) mathostelon[sic](C22) matfeloun with a rede flour(C27) materfelowne or clobwede(C31) anglice bullwede or knoppewede or hardhake(C41)

Iacinctus
Centaurea cyanus L. Cornflower, Bluebottle
gallice feverole* [corr. blaverole], anglice blowid(A6) gaverol[sic], anglice blowede(A9) gallice faverole* [corr. blaverole], blowed(A12)
carlowre [corr. carlocke?](C3) pedull, blew bochorn(C4) (aquaticus) ranculwort(C4) anglice bleu popi, langwort(C7) anglice blowode(C12) blou(d)wed(C15) a blu boiel [corr. blu boþel] vel hungerwode(C19) (aquaticus) pakelwort (agrestis) carlok or blue boyne(C21) bloweed(C23) longewort, blw popy, nydel, hongerwed, blew bodym or blw boþel(C31) nedul, blew boþul(C32) (aquatica) rankulburth[sic] or banculburth[sic] (rusticorum maior) bleu floure, nedull, bul [corr. blu] boyton(C32) (maior) blede[sic] boþel(C38) gallice brawelole[sic], anglice blowed(C39) (rusticus) anglice nedyll or blew bothell(C41)
Lit. André I, p.165 (hyacinthos) & II, pp.126f; Alph., p. 83, 22; Agnus Castus p.167, 28f. See **Blaverole**

Ibercus [for ibiscus?]
?*Althaea officinalis* L. ? Marsh Mallow
bissopeswert(A11)

Ibonfor
?*Cichorium intybus* L./*endivia* L. ? Chicory/Endive
?*Lactuca virosa* L. ? Wild Lettuce
letuse(C21)
Lit. André I, p.168 (ibonsus)

Icterites
Rosmarinus officinalis L. Rosemary
feldmadre, rosemary(C21)
Lit. André I, p.168 (icteritis) & II, p.130

Idia
Rumex aquaticus L. 'Red Dock'
Rumex sanguineus L. 'Red-Veined Dock'
reed dokke(C21)
Lit. André I, p.168 (idia)

Idia Lupina
?*Solanum nigrum* L. ? 'Black Nightshade'
anglice morel(C12)

Idrocollum
?*Anthemis cotula* L. ? Stinking Chamomile, Mayweed
anglice mayethes(C3)

Igada[?]
?*Viburnum* L. ssp.
urtica, jarus amarus, anglice hondisquelke(C4)
see **Jarus Amarus** & B III, 146

Iliaca Passio
Capsella bursa-pastoris (L.) Medic. Shepherd's Purse
passio ventris, sanguinaria .i. staunche vel bursa pastoris(A5)

Illafeos [ilapheas, ilascas, illacias]
Arctium lappa L. Burdock
clote(B3) bardane(B3)
anglice clote(C10, C12) clotre(C21) clote, burre(C21) anglice þe sede of þe bur-clote(C26) clotre(C27)
Lit. Alph., p. 85, 29 (ilapheos, illafricos) & p. 84, 10 (iancia); N184

Iluscus [for ibiscus]
?*Althaea officinalis* L. — ? Marsh Mallow
?*Malva sylvestris* L. — ? Common Mallow
holihok, wild malow(C21)

Imphium
Symphytum officinale L. — Comfrey
anglice confery(C16)

Incudus [for intubus?]
?*Conium maculatum* L. — ? Hemlock
?*Cichorium intybus* L. — ? Chicory
humbelok(C10)

Indi Dactilis
Tamarindus indica L. — Tamarind
tamaryndes(C21, C27)

Inguinale
?*Aster amellus* L. — ? Aster
lunary(C21)
Lit. André I, p.169 (inguinalis) & II, p.131

Insana
Hyoscyamus niger L. — Henbane
henbane(C27)
Lit. André I, p.169 & II, p.131

Instinctum Ortis
?*Orchis* L. ssp. — ? Orchis
ravynleeke(C21)
Lit. B II, 68 & III, 144

Intrancia
?
campworte(B4)

Intuba [intiba]
Calendula officinalis L. conf. — Marigold
w. *Cichorium intybus* L. — Chicory
gallice solsekel, anglice roddys(A1) solsekil, roddis(A3)
anglice golde(B1) gallice solsicle, anglice golde(B2)
roodyn(C10) solsecle, anglice golden(C16) gallice solsicle, anglice golde

vel redewort(C18) goold flour(C21) solsecle, marygolde, ruddes, hardhoue(C21) rudene(C24) rudune(C27)

Intuba 2
?*Conium maculatum* L. ? Hemlock
?*Cichorium intybus* L. ? Chicory
homelok(C21)
Lit. Alph., p. 86, 5 (incuba)

Inula [inile]
Allium cepa L. Onion
Allium ascalonicum L. Shallot
gallice scaloynes, anglice holleke(A1) scalons, holleke(A3)
gallice et anglice scalouns(B2) anglice scalons(B7)
gallice et anglice scalouns(C18) scalons(C22) anglice scaylon(C28)
Lit. Alph., p. 86, 10 (inule)

Inviola Alba
Viola L. ssp. Violet
whiȝt violet(C21) whyte violet(C27)

Ipomaratrum
?*Foeniculum vulgare* Mill. ? Fennel
?*Peucedanum officinale* L. ? Hog's Fennel
wild fenel(C10, C21) wild fenell(C24)
Lit. André I, p.163 (hippomarathum) & II, p.124 cf. Alph., p. 83, 4 (hues maratrum)

Ipoquistidos [ipopistidos]
Rose-gall
it growis on breris(C3) ronde ball growyng abowt a bremel(C10) a round ball groweng abowt a brymble(C21) a superfluitate þat goth oute of the breer(C24) is a rounde appull growing aboute a brembill(C24) is a superfluite of growying onþe erth þe wych berith rosam caninam(C27) fungus circa radicem rose canine, brere-appel(C36)
Lit. Alph., p. 86, 21 (ipoquistidos)

Iporis
[*Equisetum* L. ssp. Horsetail]
anglice foxglove*(C12)
Lit. André I, p.163 (hippuris) & II, p.124 See **Cauda Equina**

Iposdeos
Allium cepa L. — Onion
oynouns(B3)

Ippia Aquatica
?*Agropyron repens* (L.) Beauv. — ? Couch-Grass
?*Agropyron maritimum* (Koch & Ziz) Jansen & Wachter — ? Couch-Grass
?*Agropyron pungens* (Pers.) Roem. & Schult. — ? 'Sea Couch-Grass'
?*Lemna minor* L. — ? Duckweed, Duck's-Meat
anglice quikene(C5) anglice quichene(C6) anglice quichun(C8) anglice digmete[sic](C36)

Ippia Maior [ypia m., hispia m.]
Anagallis arvensis L. — Scarlet Pimpernel
Pimpinella saxifraga L. conf. w. Ippia minor — Burnet Saxifrage
anglice feldwurth vel heywourth .i. pimpirnole(A1 corrected) chikewort, pimpirnol(A3 corrected) pimpernele(A5)
pinpernolle, selfhele, mercroppe, wheywort(C1) chekemete(C2) pympernell, selfhele(C4) mortyn [corr. morune?](C22) anglice chikenmet, cywort(C23) morune, chikenwed(C23) chekynworte(C24) vlex, eyeworte(C24) chechunwede(C27) pympernel, mowse-ballok(C31) testiculus muris, pimpernel, selfhehe [corr. selfhele], waywort, merecrop(C32) anglice pympernolle, stiffehele [corr. selfhele?] or weywor[t]e or nerekroppe [corr. merekroppe](C41) pimpirnell oþer yworte oþer wolshele [corr. selfhele?] oþer kennyngworte(C42)
Lit. Alph., pp.184 & 196; Agnus Castus p. 230

Ippia Minor
Stellaria media (L.) Vill. conf. w. Ippia maior — Chickweed
gallice morune, anglice chikynmete(A1 corrected) gracedeu*(A2) mortine [corr. morune?], chikenemete(A3 corrected)
anglice flewort vel ciwort(B2) chikenmete(B5) chykenmett(B6)
chykynwed, chykynwort(C1) chykenmete, smalwort(C4) anglice flewort(C6) chykynmete, secundum aliquos morsus galline, fleworte(C7) vlexwort or heywort(C10) anglice flewort(C16, C18) selfhele(C21) chekunwede(C21) ankewort(C22) selfhele(C24) vlexwort, eyȝewort(C27) sedwale*(C28) flewort(C28) dukmete*(C29) chykmet or fleewort(C31) vlex anglice worte[sic], cyeworte(C40)
Lit. Alph., p.196

Ireis
Iris pseudacorus L. Yellow Flag
Iris florentina L. Florentine Iris
flowre de lyce(C1)

Ireos
Iris florentina L. Florentine Iris
wyte gladyole(C1) a white floure(C10) fane, gladiol(C21) ireos haþ a white flour, and gladiole haþ a yelu flour(C24) ireos hatz a whyte flour: anoþer is cleped gladiole and hatz a ȝelwe flour lyke saffron(C27) whytt gladyol(C38)
Lit. André I, p.171 (iris) & II, p.133

Iriana
Sonchus L. ssp. Sowthistle
thowthistell(C21)
Lit. MS ccclxxii ('id est lambrusca')

Iringus
Eryngium L. esp. *E. maritimum* L. Sea Holly
yringis, seholin(A3)
anglice thistil beneit(B1)
setacul, yringes(C10)
Lit. André p.129 (eryngium) & II, pp. 97f.; Alph. pp. 87, 7; 165, 33; 32, 27

Iris [iri]
Iris foetidissima L. Gladden, Stinking Iris
Iris florentina L. Florentine Iris
gladen(B6)
anglice flowre de lyce(C3) gladen(C21) iris hath a purpul flour(C24)

Ision [isron]
Centaurium erythraea Rafn Centaury
anglice centorie(C10) centory(C21, C24, C27)
Lit. Alph., p. 87, 21 (ision)

Isora
Arum maculatum L. Cuckoo-Pint
jare, yekesters(A3)

Itabra
?*Origanum dictamnus* L. ? Dittany of Crete

Dictamnus albus L. ? Bastard Dittany
?*Lepidium latifolium* L. ? Dittander
ditony(C21)

Itea [idea]
Salix viminalis L. Withy, Osier
anglice wythy(A1) wythy(A3)
sacherey(B1)
wilwe(C21) anglice wethy or welous(C26)
Lit. André I, p.172 & II, p.134

Iva
Plantago coronopus L. Buck's-Horn Plantain
Ajuga chamaepitys (L.) Schreb. Ground Pine
(i. media) herbe yve(C1) anglice herbe yve(C4) herbe ive(C21) (maior) more herbe ive(C21)

Jaary
[*Arum maculatum* L. Cuckoo-Pint]
ȝarowe*(C14)

Jarus [jaris]
Arum maculatum L. Cuckoo-Pint
anglice yekesters(A1) cockou-pintel(A7)
gallice morine*, anglice chikemete*(B2) ȝekespotele(B3) anglice ramose*(B5) anglice ramese*(B5) rascal, ȝekesters, cokkouispyntyll(B8)
folefot(C2) gakspatyll(C3) ȝekesters, gokkepintell(C7) anglice cokkouwespyntell(C8) gallice jare, anglice kockowpyntil(C16) dog-ballokkes(C29) pé de poleyn, volefot; ȝekestres, cokkupyntyl(C31) calvesfote(C33) jekescore*(C37)

Jarus Amarus
?
hundesquelk(C7) anglice houndiskolle(C19)
Lit. cf. B III, 146. See **Igada**

Jovis barba
Sempervivum tectorum L. Houseleek, Sengreen
gallice jubarbe, anglice silvegrene(A1) jubarbe, selvegrene(A3) gallice jubarbe(A4)
gallice jubarbe, anglice synegrene(B2) jubarbe(B3) anglice sengrene(B7)
selgrene(C21)
see **Barba Jovis**

Jucis Promeon
?*Gentiana* L. ssp. ? Baldmoney
baldmony(C21)

Judicum
?*Satureia hortensis* L. ? Savory
satureie(B3)

Jumelon Alba
?*Viola* L. ssp. ? Violet
white violet(C24)
Lit. cf. MS ccclxxiii ('iumalion .i. portulaca agrestis')

Juncus
Juncus L. ssp. Rush
Juncus maritimus Lam. Sea Rush
Cyperus longus L. Galingale
gallice junk, anglice russe(B1)
(j. magnus/rotundus aquaticus) boleryschys (j. triangl., aquaticus) flagrysches (j. marinus) seeryschys(C1) (triangularis) galingale(C21) rosses(C28) (magnus aquaticus) boleruschus (angularis) flagruschus (marinus) seeruschus (pratinus mollis) nesh ruschys(C38)

Juniperus
Juniperus communis L. Juniper
gallice junevre(A4)
gorst(B1)
gostel-tre(C2) anglice jenuper(C3) anglice gostel-tre(C12)

Jusquiamus [jusqueamus]
Hyoscyamus niger L. Henbane
gallice genyler, anglice hennebane(A1) hennebane(A3) anglice henebane(A4) anglice hennebane(A6) anglice hennebane(A8) anglice hennebane(A9, A12)
hennebane(B1) gallice geneler, anglice hennebane(B2) hennebane(B3) hennebane(B5) henbane(B6)
(j. albus) henbelle, henbane (j. rubeus, niger) henbane(C1) anglice henbane, gallice genel vel geliner, henbane or henbell(C6) chenillir, henbane(C7) gallice geneley, anglice henbane(C16) gallice chanelé, anglice henquale[sic](C39)

Labrum Veneris [labium v.]
Sonchus L. ssp. Sowthistle

Dipsacus fullonum L. Teasel
Cnicus benedictus L. Blessed Thistle
anglice southystel(A8)
anglice soutthistill(B1) anglice fouþestel(B2) souþistel(B3)
anglice fawthystyll(C3) sowthystill(C4) anglice fowthystyl(C6) fawthistel(C14) vercardun, anglice southustill(C16) vercardun, anglice a vouthestel(C18) þouþistil(C19) sowþistyl, þe sed ys callyd seynt Mary sede(C31) nosbledelis*(C33) fauþestell or oure Ladies mylke(C37) anglice seynt Marye thistill or southistill(C41)
Lit. Alph. App., p. 220; Alph., p. 89, 11; André I, p.175 & II, p.135

Labrusca [lambrusca]
Vitis vinifera L./*Vitis sylvestris* Gmel. (Wild) Vine
also ssp. of *Parthenocissus*
Sonchus oleraceus L. Sowthistle
?*Bryonia dioica* Jacq. ? Bryony
gallice vin savage(A1) vine savage(A3)
gallice vine savage, anglice wilde vine(B2)
gazer(C29) þe seed ys yclyped seynt Marie seed(C10) anglice gaser(C12) thowthistill(C21) thowthystyl, þe sede is clepyd seynt Mary sede(C27) wild vyne or wyld nepe(C29)

Lacterides [lactirida, elacteridos, lecterides]
Euphorbia lathyrus L. Caper Spurge
Daphne laureola L. Spurge Laurel
guth[c]orn(B3)
gitthorn(C2) spourge(C3) cathapucie(C8) anglice guth[c]orn(C12)
Lit. André I, p.176 (lacterida) & II, p.136; B III, 112f; Alph., p. 93, 12

Lactopos
Geum urbanum L. Wood Avens, Herb Bennet
avence(B1)
Lit. André I, p.177 (lagopus) & II, p.137; Alph., p. 89, 18

Lactuca
Lactuca virosa L./*serriola* L. Lettuce/'Prickly Lettuce'
Euphorbia L. ssp. Spurge
Sonchus oleraceus L. Milk- or Sow-Thistle
gallice letuz, anglice mylkeworth(A1) letuse, silvestris quasi domestica est, milkwort(A3)
runcele, soutthestil(B1) (silvatica) letuse dé boys(B1) gallice letuse, anglice milkewort(B2) letuse, slepwort(B8)

(domestica, ortulana) letuse, slepwort (leporica) hare-letuse, hare-þystyl(C1) endyve(C2) anglice letus(C3) (domestica) slepewort, letuse (silvatica) wode-letuse (agrestis) wilde letuse (leporina) hare-thistill or hare-letuse(C4) gallice letuse, anglice myl[ke]wort, slepeworte(C6) (agrestis) mylk-thistell(C7) letewes(C14) gallice letus, anglice milkwort, slepwort(C16) (domestica) letuse, cleywort [corr. slepwort](C21) (leporica) hare-thistill(C21) (leporica) hare-dystyll(C32)
Lit. B III, 154 & 239

Lactucella
?*Sonchus oleraceus* L. ? Milk- or Sow-Thistle
anglice sowthistill(C16)
Lit. cf. André I, p.177 (lactuscula) & II, p.136

Ladanum
?*Phyllitis scolopendrium* (L.) Newm. ? Hart's-Tongue Fern
hertystong, cerela[n]g(C1)

Lanceolata [lancealata]
Plantago lanceolata L. Ribwort
gallice launcely(A5) anglice rybwrt(A6) ribbeuort(A8) anglice ribbewort(A9) anglice ribwort(A10) ribworth(A12)
(aquatica) anglice sperewort(B1) gallice lancelé, anglice ribewort(B2) lancelé, ribwort(B3) rybworte, lancelé(B4) ribwort(B5) lancelé, rybwort(B8)
(maior, flammula aquatica) sperworte (minor, plantago minor) rybwort(C1) anglice rybeworte(C3) rybbe vel rybbewort(C4) lancelet, anglice ribwort(C12) ribbeworte, launceolé(C24) (aquatica) sperewort(C28) (maior) sperewort (minor) rybwort(C38) romani vocant eam lancelé, anglici lovesche*(C43)

Lapillus
Rumex acetosa L./*acetosella* L. Common Sorrel/Sheep's Sorrel
Oxalis acetosella L. Wood Sorrel
anglice sowredoke(C3)

Lapis Judaicus
?
sporge(C21)

Lappa
Arctium lappa L. Burdock
gallice clot vel diswort(B1)

clote(B3) burre(B4) clote(B5)
(l. maior, agrestis) burre, clyþe(C1)

Lappa Eversa [l. inversa, l. aversa]
Arctium lappa L. Burdock
Arctium minus Bernh. Lesser Burdock
gallice gletuner(A2) clote vel glettiner(A3) gallice gletenere, wlvedok(A6) gallice gletenere(A8) gallice gletoyner(A9) gallice gleytener(A12)
gallice glestron(B9)
(l. minor, aquatica, aversa, inversa) clote(C1) anglice burdoke(C3) gallice cardoun, anglice a clote(C6) anglice thistell(C15) anglice clot vel burr(C16) clete, burre(C21) dystil(C32)

Lappana
Arctium lappa L. Burdock
anglice schalgras(C16)

Lappates
?*Brassica oleracea* L. ? Cabbage
cabbochewort(C21)

Lappatium [lappacium]
Rumex L. ssp. esp. *sanguineus* L. Dock esp. 'Red-Veined Dock'
Arctium lappa L. Burdock
gallice parele, anglice docke(A1) gallice parel, anglice dok(A4) clote(A11)
rede docke, ruge parell(B4)
clete(C3) þer beth .iii. maner docke: scharpe-levyd, round-levyd, longe-levyd(C30) (maius agreste) longge dok (minus aquaticum) round waterdok(C38)

Lappatium Acutum
Rumex aquaticus L./*sanguineus* L. 'Red Dock'/'Red-Veined Dock'
Rumex hydrolapathum Huds. 'Great Water Dock'
anglice red doke(A2, A12) ruge parelle, rede dokke(A3) anglice red doc(A6) rede docke(A8) anglice rede doke(A9)
gallice rouge parele, anglice rede docke(B2) anglice rede dock, gallice parel(B9)
(l. maius, longum, acutum agreste) longe docke (magis, acutum, longum aquaticum) long waterdok(C1) rede doke(C3) litel dok(C10) smal dok(C21) (agreste) longge dok (aquaticum) longge waterdok(C38) red clote [corr. doce?](C39)

Lappatium Rotundum
Rumex L. ssp. esp. *obtusifolius* L. Dock esp. 'Broad-Leaved Dock'
Arctium lappa L. Burdock
gallice parele(A2), rounde parelle, rounde dokke, pobel(A3) vocatur anglice vel hibernice pobel(A5) anglice clete(A6) clete(A8) anglice clote(A9, A12)
anglice wite dokk(B1) clote(B3) anglice clete(B9) (l. minus, rotundum aquaticum) rond waterdok(C1) mechedok(C10) muchedocke(C11) mochedokk(C21) þe moch dok(C24) mechedok(C27) (agreste) rownde dok(C38)

Laurencia
?*Lavandula officinalis* Chaix ? Lavender
lavendre(C21) lavendyre(C27)

Laureola [lauriola]
Daphne laureola L. Spurge Laurel
Laurus nobilis L. Laurel
Daphne mezereum L. Mezereon
gallice laureole(A1) laureol(A3) gallice laureole(A4)
anglice et gallice lauriole(B1) gallice et anglice laureole(B2) laureole(B3) lorer(B5)
lauriol(C1)

Laurus
Laurus nobilis L. Laurel, Bay
gallice lorer(A1) lorer(A3) gallice lorer(A4)
lorer(B1) lorele(B3)
lorel-tre(C24)

Lauscea Rustica
?*Cannabis sativa* L. ? Hemp
hemp(C21)

Lavendula
Lavandula officinalis Chaix Lavender
Lavandula latifolia L. French Lavender
gallice lavendre(A1) lavendre(A2) lavendre(A3) gallice lavendre(A6, A9, A12)
lavendre(B1) gallice et anglice lavendre(B2) lavendule(B3) lavendyr(B8)
lavendyr(C1) lavendure(C24)

Leconta
Lactuca virosa L. Lettuce
letus, slepworte(B4)

Lectarcula
?*Hyoscyamus niger* L. ? Henbane
henbane(C21)

Lectorica [lectoria, lentorica]
Corylus avellana L. Hazel
fylberd(C10) fylbordys(C11) fylberdis(C21) filberdys(C24)
fylberdys(C27)

Lentes [lendes, lens]
Lens esculenta Moench Lentil
Vicia L. ssp. Vetch
anglice fichez(B7)
gallice et anglice tilles(C5) anglice tylles(C6) ffeche(C21)

Lenticula Aquatica
Lemna L. ssp. esp. *L. minor* L. Duckweed, Duck's-Meat
gallice lentyl, anglice enede-met(A1) lentil de ewe, enede-met(A3)
gallice lentile de ewe, anglice henede-mete(A4)
anglice henede-mete(B2)
(l. maior) chyche (macro l.) chyches(C1) gallice lentyl d'ewe, anglice ende-mete or duckmete(C6) lentildew, anglice ende-mete(C10) ende-mete, lentildowe(C10) lentile(C14) lentildue(C21) lentil de ewe(C22) lentildow(C24) enede-mete, lentildw(C27)

Lenticulum
?*Lens esculenta* Moench ? Lentil
?*Vicia* L. ssp. ? Vetch
tille, wodebynde*(C21)

Lentigo
Lemna minor L. Duckweed, Duck's-Meat
anglice enede[-mete](A8)
chikenmete(B3)
doukmete, ende-schede[?](C1) chekemete(C2) dokemete(C2) anglice dokemette(C3) dukkemede(C4) anglice chekmete(C12) doumete(C14) hertesgras*(C15) dokesmete(C21) dokmete, neþergrece*(C28) ened-gras(C32)

Lentiscus
[*Pistacia lentiscus* L. Mastic]
prevet*(B3)
anglice byrche*(C6, C33, C37) lentiscus is a tre, yt schetythe owte oyle(C40)
Lit. C28 = alba spina

Lentos
Lens esculenta Moench Lentil
Vicia L. ssp. Vetch
anglice tylles(C6)

Leontopodion [lentopodion, letipodion]
Alchemilla vulgaris L. Lady's Mantle
Helleborus niger L. Christmas Rose
pé de lioun(B1)
more pedilion(C21)
Lit. André I, pp.183f & II, p.142

Leporina [leporica]
Orchis L. ssp. Orchis
?*Endymion non-scriptus* (L.) Garcke ? Bluebell, Wild Hyacinth
ravenleek(C21) harebell, croweleke(C21)
Lit. B III, 144; André I, p.184 & II, p.142; N188

Leptafillos
?*Conium maculatum* L. ? Hemlock
anglice hemlok(C16)
Lit. B III, 153. cf. André I, p.184 & II, p.142

Leucopiper
Eruca sativa Mill. 'White Pepper'
gallice blaunk peyvre, anglice wyt piper(A1) wite pepir(A3)

Levisticum [luvesticum, levistica, levisticus, levesticus, leviscus, leviscum, livesticum]
Levisticum officinale Koch Lovage
Ligusticum scoticum L. Scotch Lovage
gallice luvache(A1) lovache(A3)
anglice lofache(B1) gallice levistiche, anglice loveache(B2)
lovehache(B3) lovage(B4) lovach(B5) lovestiche, loveache(B8) anglice loefache(B9)

loveache (l. agreste, minus) lasse loveache(C1) lovestiche, anglice loveache(C10) loveache, lovestych(C27)

Levisticus Agrestis
?Ligusticum scoticum L. ? Scotch Lovage
wilde lovach(A7)
anglice clofþunke*(C12)

Libanotidos [libanitis]
Rosmarinus officinalis L. Rosemary
rosemarine(C11) anglice nedilwort(C16) rosmary(C21) rosemari, feld-mader(C21) rosmaryne(C27)
Lit. André I, p.185 (libanotis) & II, p.144

Libanus
Rosmarinus officinalis L. Rosemary
anglice rosmaryn(C16)
Lit. André I, p.185 (libanotis) & II, p.144

Liburnea
Agrimonia eupatoria L. Agrimony
egremony or goseclite(C21)
Lit. André I, p.186 (liburnia) & II, p.144
Licium
Lonicera periclymenum L./ *caprifolium* L. Honeysuckle/'Perfoliate Honeysuckle'
wodebynde(C27)

Lignum Crucis
Viscum album L. Misteltoe
mystel(C21) anglice oke-mystellis(C26)

Lignum Paradisi
Aloë L. ssp. Aloes
lignum aloes, anglice a tre of paradise(C26)

Ligustrum [lugustrum, ligustum, lugustum]
Primula veris L. Cowslip
Lonicera periclymenum L. Honeysuckle
Convolvulus arvensis L. also var. applied Bindweed
lili savage, wild lilie(A3)
.iii. maner de lis, gallice liz savage(B2)

cowsclippe, cowschyppe(C1) primrose(C4) secundum quosdam dicitur waturlylye(C6) secundum quosdam anglice honysok(C6) wild lylye(C10) primerole(C14) tres maners de liz, gallice liz savage, anglice cowsclope(C16) primerose(C21) wild leli(C21) waterlilie(C23) wilde lylly, primerose(C24) wodbynd floures(C26) malow, cowslop, honysokell(C28) honysokyll(C29)
Lit. B III, 147; MS ccccxcii ('volubilis maior . . . caprefolium')

Lilifagus [lilifagi]
Salvia officinalis L. — Sage
Teucrium scorodonia L. — Wood Sage
gallice sauge(A1) sauge(A3) wilde lilie*(A5)
gallice et anglice sauge(B2)
ambrosie, wilde sawge(C2) hyndhale, wild saug(C10) anglice ambrose(C12) ambrose, hindehale(C21)
Lit. André I, p.124 (elelisphacos) & II, p. 93; N189

Lilium
Lilium L. ssp. — Lily
gallice liz, anglice lilie(A1) liz, lilie(A3)
gallice liz, anglice lilie(B1) gallice liz, anglice lilie(B2)
(l. domesticum, ortulanum) lylye (aquaticum) waterlilie(C1)

Limphea Aquatica [limpha a.]
Nymphaea L. ssp. — Water-Lily
waterlilie(A3) anglice docceflure(A8)
waterlilie(B3)
waterlylye(C10) edoke(C15) water lily(C24)

Linaria
Linaria vulgaris Mill. — Toadflax
Cuscuta epilinum Weihe. — 'Flax Dodder'
linarie(B1)
lynarie(C1) anglice dodyr(C3) wylde flex(C5) flexwort(C7) wilde flax(C28)

Lingua Agnina [l. agni]
Plantago major L./*media* L. — 'Great Plantain'/'Hoary Plantain'
Plantago lanceolata L. — Ribwort
weybrede(A3)
gallice plaunteigne, anglice weybrede(B2)

lombystonge(C1) weybrede, lambystung(C4) lambtong(C14) lammestung(C21) planteyn(C21) anglice rybwort or lanciel(C26)

Lingua Anserina
Hieracium pilosella L. Mouse-Ear Hawkweed
mouser(C7) anglice mousere or gose-tonge(C26)

Lingua Arietis
Plantago major L. 'Great Plantain'
planteyn(C7) gallice plantayn(C16) playnteyn, weybreed(C21)

Lingua Avis
Stellaria holostea L. Stitchwort
Fraxinus excelsior L. Ash-Key ('semen fraxini')
anglice sticheworth, gallice pigle(A1) anglice pigle(A2) semence de frenne(A3) anglice stichewort, gallice pigle(A3) stichewort, pigil, briddestounge(A3) gallice pigle, stychwort(A6) anglice pigel(A8) anglice pigle(A9) pigle(A12)
anglice stichewort(B1) gallice la semence de freyne, anglice stichewort(B2) gallice pigle, anglice briddestonge(B2) pigle(B3) stechewort, wesseltonge, pygyle(B4) pigle(B5) pegle(B5)
anglice pygyll(C3) pygull, stichewort vel byrdestung(C4) semence de frene, aliquando herba .i. pigle, stychwort(C22) semence de freyne, herba stichewort(C23) anglice aysh-keyes(C26)
Lit. Alph., p.103, 21

Lingua Bovina [l. bovis]
Anchusa officinalis L./ *arvensis* (L.) Bieb. Alkanet/Bugloss
?*Picris echioides* L. ? 'Bristly Ox-Tongue'
?*Borago officinalis* L. ? Borage
ruþerestunge(A3)
gallice lang de bove .i. hoxtong(B1) anglice reþerestonge(B2)
longe de besfe (minor, agreste) wylde long de bef(C1) oxtung or lang de bef(C4) anglice oxtonge, wild burage(C16) (agrestis) wild lang de bef(C21) retherestunge(C23) (agrestis) wylde lange de bef(C27) bugyll* or oxtonge(C29)

Lingua Canina [l. canis]
Cynoglossum officinale L. Hound's-Tongue
anglice hundestong(B1)
hounstong(C1) houndestong(C4) stikelwort(C27)

Lingua Caprina
?*Coronopus squamatus* (Forsk.) Aschers. — ? Swine-Cress
gootestung(C21)
Lit. Alph., p.103, 15

Lingua Cervi [lingua cervina]
Phyllitis scolopendrium (L.) Newm. — Hart's-Tongue Fern
cerelaunge(A9)
cerlange(B3) cerlaunge(B5)
herstystone [corr. hertystong](C1)

Lingua Cervina
?*Echium vulgare* L. — ? Viper's Bugloss
?*Plantago coronopus* L. — ? Buck's-Horn Plantain
?*Coronopus squamatus* (Forsk.) Aschers. — ? Swine-Cress
swynesgresse, bukkhorn(C21)

Lingua Hircina [l. hirci]
Coronopus squamatus (Forsk.) Aschers. — Swine-Cress
Plantago lanceolata L./ *coronopus* L. — Ribwort, Buck's-Horn Plantain
?*Phyllitis scolopendrium* (L.) Newm. — ? Hart's-Tongue Fern
swynysschesse[sic], bokhorne(C1) anglice bochorn, swynesgrasse(C4) gotistung(C7) gotestonge(C8) bukkeshorn, swyneskerse(C32) hertistunge(C35) bokishorn(C38) gallice de langue clevere [corr. langue de chevre], anglice ribbewort(C39)
Lit. Alph., p.103, 15

Lingua Passerina [l. passeris]
Polygonum aviculare L. — Knotgrass
sperwetunge, cnotgras, swynysgras(A3)
swynesgrasse(C4) speretonge vel stychesgras(C6) anglice cnottedgras(C12) stichegresse(C21)

Lingua Serpentis
Lysimachia nemorum L./*vulgaris* L. — 'Yellow Pimpernel'/'Yellow Loosestrife'
Gagea lutea (L.) Ker-Gawl. — 'Yellow Star-of-Bethlehem'
Ophioglossum vulgatum L. — Adder's Tongue
eddyrtonge (minor, aquatica) lasse eddertonge(C1) addertung(C4) neddretung(C21)

Lingua Silvana
?*Anchusa officinalis* L./ *arvensis* (L.) Bieb. — ? Alkanet/Bugloss
lang de bef(C21)

Lingua Ursina
Primula veris L. — Cowslip
cowsloppe(C8)

Linozostis [linachides, linochites]
Mercurialis perennis L./*annua* L. — Dog's Mercury/'Annual Mercury'
Chenopodium bonus-henricus L. — All-Good, Mercury, 'Smearwort'
gallice mercuri, anglice smerewurth(A1) mercurial, smerwort(A3) kerlekes*(A11)
gallice mercurial, anglice smerewort(B2) mercurie(B3) mercurial, anglice smerewort(B7)
mercurie(C2) mercureal(C10) smeredok or mercurie(C14) mercuri, papwort, more smerwort(C21) mercuriell(C24)
Lit. André I, p.188 (linozostis) & II, p.146

Linum [linium]
Linum usitatissimum L. — Flax
flex(A3)
(maius, domesticum) flex, lynen (minus, agreste) wylde flex(C1) lyne or flex(C4)

Lippilli [luppuli]
Melilotus officinalis (L.) Pall. — ? 'Common Melilot'
Trifolium ornithopodioides L./ *pratense* L. — ? 'Birdsfoot Fenugreek'/'Red Clover'
cokkis-heveden(B1)
cakkeshevd(C28)

Liquiricia [glicoricia, gliceria, liquericia]
Glycyrrhiza glabra L. — Liquorice
licoriz(A1)

Lithodemonis
?*Nigella* L. ssp. — ? 'Fennel-Flower'
geet(C21)

Lithospermon [licosperma]
Lithospermon officinale L. Gromwell
gallice la semence de la saxifrage(A1) la semence de saxifrage(A3)
sundcorn(B1) gallice la semence de saxifragie(B2)
gallice semense de saxifrage(C16) saxifrage-seed(C21) semence de saxfragi(C22)
Lit. Alph., p. 99, 5; Alph., App. p. 219; B III, 222; N189

Lolium [lollium]
Agrostemma githago L. Corn Cockle
Lolium temulentum L. Darnel
gallice neel, anglice cokyl(A1) nel, cokkel(A3) anglice kockel(A6)
anglice cockel(A8) anglice kokel(A9) gallice nel, anglice kokel(A12)
gallice nelle, anglice cokil(B1) gallice neel, anglice coc(B2) cokkel(B3)
kockel(B5)
cokkyl, nele, wyld flos campi(C1) anglice popyll(C3) cokkill, popy, wild sanegre(C21) anglice kokkyll or wylde singrene [corr. sinagre](C41)
wyld sanagre(ne) oþer cokkyll(C42)

Lumbricus
?*Peucedanum officinale* L. ? Hog's Fennel, Sulphur-Weed
wormsede(C21)

Lunaria
?*Botrychium lunaria* (L.) Sw. ? Moonwort
?*Aster amellus* L. ? Aster
?*Lunaria annua* L. ? Honesty
(minor) moneherbe(C1) lunary(C4, C21) purgiþ þe colour þe wiche bereþ a white flour(C10) purgeth þe colre þe whyche hatz a bleu flour(C27)

Lunnosus
? ?
centorie(B3)

Lupina
Atropa bella-donna L. Deadly Nightshade
gret morele(B3)
anglice gret morel(C12)

Lupinus
Lupinus L. ssp. Lupin
gallice lupins(A1) lupins(A3)

(maior, agrestis) lupyn (minor, montanus) lasse lupyn(C1) anglice lupyns(C3)

Lupi-Pecten
Carduus acanthoides L. 'Welted Thistle'
Dipsacus sylvestris Huds. Wild Teasel
tasyll(C21)

Lupus [lupulus]
Rumex acetosa L./*acetosella* L. Common Sorrel/Sheep's Sorrel
Oxalis acetosella L. Wood Sorrel
surel(C8) swredok(C16)

Luteum [lutea, lucia]
Reseda luteola L. Dyer's Rocket, Weld
wolde(C7)
Lit. André I, p.191 & II, p.149

Macedonia [macedonium, macedonicum, macedonica]
Smyrnium olusatrum L. Alexanders, Horse-Parsley
gallice alisandre(A2) alisandre(A3) alexaundre(A6) stanmarche anglice alisander(A9) gallice alisandre, anglice stanemarch(A12)
alisaundre(B3)
alisaunder, stanmerche(C1) anglice alysaunder-seede(C6)

Macis
Myristica fragrans L. Nutmeg
.i.maner de noys(B2)
anglice the blome of þe flour of notemyg(C26)

Macofilon
Iris L. ssp. Iris
gladyn, bladyn(C21)
Lit. André I, p.194 (machaerophyllon) & II p.151

Macropiper
Piper longum L. Long Pepper
lung pei[v]re(A3)
lung peivere(B1) gallice lunge peyvre, anglice long peper(B2)

Maculata
Trifolium L. ssp. Clover
clavergras(A3)

anglice claveregras(B2)
anglice clove[r]gras(C6) clove[r]gras(C8) þre-levede gras(C10) trifoile(C21) anglice clavergras(C22) þe .iii. leved grasse(C24) clavergras(C36)

Magedan [magedena]
Rubia tinctorum L. — Madder
warence, madyr(A3)
anglice made[r](C32)

Majora [majorana]
Origanum vulgare L. — Marjoram
Majorana hortensis Moench — Sweet Marjoram
gallice majoraunce(B2)
gallice majoraunce, anglice majorane(C16) gallice majorance, anglice majorane(C18)

Majoralia
[= majorana]
wallwort*(C34) wallewort*(C37)

Majorana [magirona]
Origanum vulgare L. — Marjoram
Majorana hortensis Moench — Sweet Marjoram
Anthemis cotula L. — Stinking Chamomile
?*Sambucus ebulus* L. — ? Danewort, Dwarf Elder
anglice hundefenyle(A2) hundefenyl(A3) gallice majorane(A4)
anglice sambuk, eldirne*(B1) maithe(B2)
majoran (minor) lasse majoran(C1) majorane(C2) muchewort(C10) methewort(C11) anglice majorane vel mathys(C12) maythes(C14) mugwort(C21) merchworte(C24) mugwort(C27) anglice majeron(C30) magiron(C42)
Lit. B III, 68 & 80

Mala Citonia [citonica]
Cydonia oblonga Mill. — Quince
gallice coyns(A2)
anglice quinsys(C30)

Mala Coctana
Cydonia oblonga Mill. — Quince
coynz(A3)

anglice coyntz(B2) anglice coyns(B7)
qwyns(C12)

Mala Granata [malum granatum]
Punica granatum L. Pomegranate
gallice pome gernete(A4)
anglice pome garnat(C6) gallice pome garnette(C20)

Mala Matiana
Malus sylvestris L. Crab Apple
anglice wode-aples .i. wode-grobbis(A1) pumme dé boys,
wode-crobbis(A3) anglice wode-grobbis, gallice poumes savages(A4)
gallice coyns(A5)
gallice poume dé bois(B2) crabbes(B3)
crabbes(C2) wode-crabbes(C10) gallice pome dé boys(C20)
wod-apples(C30) soure crabbes(C37)

Malacarpon
Rubia tinctorum L. Madder
reed madre(C21)
Lit. cf. André I, p. 205 (melocarpon) & II, p.158

Malaspermon
Agrostemma githago L. Corn Cockle
Lolium temulentum L. Darnel
cokkel(B3)

Malgina
?*Galium aparine* L. ? Cleavers
hairef(C21)

Malochinagria [molachinagria]
Asperula odorata L. Sweet Woodruff
woderove(B3)
anglice wodrove(C12)
Lit. B III, 267f. cf. André I, p. 210 (moloche agria) & II, p.163

Malum Avellanum
Corylus avellana L. Hazel-Nut
smal notte(A3)

Malum Macedonicum
Malus sylvestris L. Crab Apple
gallice pome dé bois(C6) pomme dé boys, crabbe(C6)

Malum Paradisi
Musa paradisiaca L. Plantain
Musa sapientium L. Banana
anglice paradise-appel(C26)

Malum Punicum
Punica granatum L. Pomegranate
gallice pome gernete(A1) pumme gernette(A3)
gallice poume gernet(B2)
pomme garnate(C22) pome garnet(C27)

Malum Storacis
?*Aristolochia rotunda* L. ? Aristolochia, Birthwort
astrologe, galingale, meke(C21) anglice holywort or round arstrologie or galyngale, meke(C26)
Lit. Alph., p.107, 19

Malum Terre
Conopodium majus (Gouan) Loret Pignut, Earthnut
Cyclamen hederifolium Ait. Sowbread
anglice erthyenote(A1) erþenotte(A3) herthn[u]te(A8)
anglice erþenote(B1) ertnote(B2) erthnote(B3) anglice yornoth(B9)
(minus) lase erþenote (maius) þe more erþe n[ot]e(C1) erthnote, gallice noice de terre(C7) a mandrake appill(C21) erthnote, colueran(C22) a mandragge happil(C27) peletre*(C28) harnotte(C29)

Malva
Althaea officinalis L. Marsh Mallow
mauve .i. hokkys(A3)
anglice malves(B1) gallice wymauve, anglice mersmalue(B2)
hokke(C1) wymalfe, holy malfe(C7)

Malva Agrestis
Malva L. ssp./*Althaea* L. ssp. Mallow
smale hockes(A1) la petite mave, smal hokkys(A3) gallice mauve, anglice smal hokes(A4)
la petite malve(B2)
feld-malue(C1)

Malva Crispa
?*Malva* L. ssp. esp. *Malva crispa* L. ? Mallow esp. 'Crisped Mallow'
?*Malva moschata* L. ? Musk Mallow
?*Geranium molle* L./ *columbinum* L. ? 'Dove's-Foot Cranesbill'/ 'Long-Stalked Cranesbill'
mawort, sparge(C21)
Lit. Agnus Castus, p.247f; cf. Alph., p.110,32

Malva Viscus [malviscus]
Althaea officinalis L. Marsh Mallow
?*Anthemis cotula* L. ? Mayweed, Stinking Chamomile
gallice wymauve, anglice mersmauve(A1) wymawe, mersmaue(A3)
holihokke(B3)
white malves(C14) holihok, wild malwe, mayht, maidwede or houndisfenkel(C21) holy malow(C29)

Mamus
Boswellia thurifera L. Frankincense
franc encens(A3)

Mandragora [mandragore]
Mandragora officinarum L. Mandrake
mandragre(A3)
mandragre(B1) gallice et anglice mandrage(B2) mandrake(B3,B4)
(m. femina) femel mandrake(C1) gallice et anglice mandrak(C18)
mandrag, erth-apple(C28)
Lit. Alph., p.109, 32; André I, p.199 & II, pp.153f.

Manichos
Solanum nigrum L. 'Black Nightshade'
litil morel(C21)
Lit. B III, 98f; André I, p.198 (manicon) & II, p.154; Alph., p.176

Manumartis
Potentilla reptans L. Cinquefoil
quinfoile(C21)
Lit. André I, p. 209 (manus Martis) & II, p.154

Maratrum [marcitrum]
Foeniculum vulgare Mill. Fennel
semence de fenoil(A3)
gallice semence de fenoil(B2) fenkelsede(B4)

(m. nigrum) red fynel(C1) fenelseed(C10) gallice semense de fenoyle(C16) fenelseed(C21) sement de fenel(C22) (album) whit fenell (rubium) red fenell(C38)

Margarita
?*Artemisia vulgaris* L. ? Mugwort
mugwort(C19)

Maronion
Centaurium erythraea Rafn Common Centaury
more centory, erthgalle(C21)
Lit. André I, p. 200 (maronion) & II, p.155

Marrubium [maurubium]
Marrubium vulgare L. White Horehound
Ballota nigra L. ssp. *foetida* Hayek Black Horehound
?*Rubia tinctorum* L. ? Madder
Lamium album L. White Dead-Nettle
gallice marole, anglice horehune vel horehund(A1) marole, horhunne(A3) gallice marole(A4) anglice horehune(A5) anglice horhoune vel hundesnetele(A6) anglice harhoune vel hundenebel [corr. -netel](A8) anglice horhoune(A9) horhune(A11) horhune vel blinde-netle(A12)
anglice horshove*(B1) gallice maroil, anglice horhoune(B2) horhoune(B3) herhone(B4) (m. rubeum) mader(B5) gallice mariol, anglice harhoune(C26)
horehound(C1) anglice harehoffe*(C3) gallice marchull, anglice horhound(C6) (m. album) mousere*, horehound(C7) marioill(C10) maroel, rodelet(C16) moriol, horehound(C22) anglice mawrolle or horehounde(C41)
Lit. Alph., p.111, 6

Marsilium [marcillum, martillium, martillum, marcilium]
Lupinus L. ssp. esp. *L. albus* L. Lupin
lupins(A3)
gallice lupine(B2)
teselwort(C7) anglice lupine(C10) cetelworth(C15) gallice lupin(C16) lupines(C21) lupine(C22) lupynes(C24)

Martisum
?*Sambucus ebulus* L. ? Danewort
walwort(C21)

Masmatrum [for masmaratrum]
Peucedanum officinale L. Hog's Fennel
houndfenel(C10) houndesfenel(C11) houndefenel(C21)
Lit. N191

Mater Filius [for matrisilva?]
?*Asperula odorata* L. ? Sweet Woodruff
?*Asphodelus ramosus* L. ? Asphodel
anglice woderove(C19)

Mater Herbarum
Artemisia vulgaris L. Mugwort
anglice mogwort, gallice armoyse(A8) gallice armoyse, anglice mugwort(A9) gallice armoyse, anglice mogworth(A12)
mogwort(B3)
mogwort(C1) gallice armoyse(C3)

Mater Silvarum [mater silve]
?*Clematis vitalba* L. ? Traveller's Joy
Lonicera periclymenum L. Honeysuckle
?*Convolvulus arvensis* L. ? Bindweed
?*Glaucium flavum* Crantz ? 'Yellow Horned-Poppy'
cheverfoil(B3)
wodbynde(C14) woodbynd, memithe, wild celidony(C21)
Lit. N191

Mater Salvie
?*Salvia pratensis* L. ? 'Meadow Clary'
clarrey(C21)

Materna [materana]
Lonicera periclymenum L. Honeysuckle
Artemisia vulgaris L. Mugwort
?*Cicuta virosa* L. ? Cowbane
?*Glaucium flavum* Crantz ? 'Yellow Horned-Poppy'
?*Anthriscus sylvestris* (L.) Hoffm. ? Cow Parsley, Keck
anglice wodebyne(C16) mater herbarum, matriceria, kekkis, memithe, wodewhistil(C21)
Lit. B III, 259 (wodewistle)

Matricaria
Artemisia vulgaris L. Mugwort

mogwort, ermoyse(C7) muggewort(C10) anglice mogwort(C12) mugwort(C21) matricary(C26)

Matrisilva
Lonicera periclymenum L. — Honeysuckle
wodebynde(C7, C11, C24)

Matrona
?*Artemisia vulgaris* L. — ? Mugwort
fethirfue(C21)
Lit. André I, p. 202 & II, p.156

Matulea
?*Centaurea nigra* L. — ? 'Lesser Knapweed', Hardheads
scariol*, hardhak(C21)

Maturella [maturelle, maturielle]
Valeriana officinalis L. — Valerian
valeriane(B3)
valerian(C10) wodrove*(C12) valerian(C21)
Lit. N192

Maura [meura]
Peucedanum officinale L. — Hog's Fennel
?*Ononis repens* L. — ? Restharrow
?*Scrophularia nodosa* L. — ? Figwort
restebouf.i. kammok(B1)
cammok(C28)
Lit. André I, p. 202 & II, p.156

Meaton
Papaver L. spp. — Poppy
wyld chesseboll(C21)
Lit. cf. André I, p. 202 (mecon) & II, p.156

Mecos
Papaver L. ssp. esp. *P. somniferum* L. — Poppy
whiȝt chesseboll(C21)
Lit. André I, p. 202 & II, p.156

Medera [modera]
Anagallis arvensis L. Scarlet Pimpernel
pympernel, waywort, morecrump [= morecrop](C21)

Medica Sancte Marie [m. herba Marie]
Delphinium ambiguum L. Larkspur
larkispore, ladiis-gresse(C21) ladys-laddere(C32)

Meditullium
Helleborus L. ssp. Hellebore
Veratrum album L. White Hellebore
lungwort, elebre(C21)

Megrea
Polygonum aviculare L. Knotgrass
centinodi(C21)

Melancium [melantum, melancia]
Nigella arvensis L./*sativa* L. 'Black Cumin', 'Fennel Flower'
alisaundre*(B3) anglice cokkul(B7)
anglice cockull(C6) cokel(C10) anglice cockil(C19) cokkill, popy, [corr. popyl?] neele, wild sanegre(C21) gith(C21) childe [corr. wilde] corn(C28)
Lit. André I, p. 204 (melant(h)ion) & II, p.157

Melandria [melande, melaundria, molandria, melandea]
Melissa officinalis L. Balm
berewort(A3)
anglice berewort(B2)
berewort(C1) berewort(C2) anglice berewort(C6) berwort(C8) b(r)erewort(C10) b(r)eerwort(C11) anglice berewort(C12) anglice berewort(C18, C27)

Melangia Alba/Rubea
?
melizana alba/nigra, melizane(C1) melizane(C38)

Melanpiper
Piper nigrum L. Black Pepper
neir pei[v]re(A3)

Melantia
?*Cichorium intybus* L. ? Chicory, Wild Succory
anglice chiccorin(B1)

Meleuma
?*Melissa officinalis* L. ? Balm
medewort(C28)

Melissa
Melissa officinalis L. Balm
gallice pigmentarie(A1) anglice medewort(A2) curungen(A2) pigmentarie(A3) anglice medwourt(A6) anglice medewort(A8) anglice medewort(A9) anglice maydewort(A12)
medworte(B3) medwort(B4)
melago, anglice pymentarie(C1) bawme(C2) anglice maydyngyrs(C3) anglice medeworth(C6) melose, medwort(C7) balme(C14) reynette, anglice medewort(C20) melago, baume, pigmentary(C21) medesote vel borwort[sic](C28) maydengres(C28) nepita, anglice nepte*(C29) maydenworte(C37)

Mellilotum [melliloqus, melilota]
Melilotus Mill. var. Melilots & Clovers
Trifolium L. ssp.
gallice mellilote, anglice hunisoke vel red clavre(A1) [= Trifolium pratense] mellilote, honysoke, red clavre(A3) anglice hunysouke(A6) anglice honisuccheles(A8) anglice honisokel(A9) anglice honisokle(A12)
sukkel de serfoile(B1) gallice mellilote vel coroune de reyne, anglice honisoucle(B2) (m. rubeum) honisoukes(B3) mellilot, on of þe .iii. levid gras(B8) (rubeum, medium) red honysowke(C1) thre-levyd gras, cockhedis(C7) anglice mellilot vel honysokyl(C12) hertclaver(C15) gallice millote[sic], vel coroune de regine, anglice honisouke(C16) gallice mellilote vel coroune de reyne, anglice houseleke*[sic](C18) mellilote, honysuckle(C21) trifel, .iii. levid gerse(C28) medwort(C35)
Lit. André I, p. 204 & II, p.158

Mellipopo
?*Cucumis melo* L. ? Melon
?*Bryonia dioica* Jacq. ? Bryony
wilde nep(B3)
Lit. André I, p. 205 (melopepo) & II, p.159

Melochia [melochie]
Althaea officinalis L. Mallow

hokkes(B3)
Lit. André I, p. 210 (moloche) & II, p.163

Melones
Cucumis melo L. — Melon
gallice meluns(A5)

Melus
Mentha pulegium L. — Penny-Royal
Thymus serpyllum L. — Sweet Marjoram
Origanum vulgare L. — Marjoram
anglice broþerwort(C12)
Lit. Alph., p.113, 5

Memita
Glaucium flavum Crantz — 'Yellow Horned-Poppy'
celidoyne sauvage(A7)
.i. maner de celidone(B2)
celydoyne de mere(C7) celydonye de mer(C12) anglice celydoyne(C22)
Lit. André I, p. 206 (memitha); Alph., p.114, 31 & 33 & p. 36, 7

Men
?*Anethum graveolens* L. — ? Dill
?*Meum athamanticum* Jacq. — ? Spignel, Meu, Baldmoney
?*Peucedanum palustre* (L.) Moench — ? Hog's Fennel, Milk Parsley
dile(A3)
anglice gromyle*(C22) is an herb lik hondesfen [sic], but it is mor croked(C24) (semen men) anglice wyld anet seed(C26) is an herb lyke hounde-fenel, bote it is more crokyd(C27) similis est feniculo sed tortuosa(C37)
Lit. Alph., p.116, 29

Menelata
Chrysanthemum segetum L. — Corn Marigold
?*Convolvulus arvensis* L. — ? Bindweed, Cornbine
goldesboþom(C10) goldysbothun(C21) bodwine(C23)
goldysbothun(C27)
Lit. Alph., p.112, 5

Menphea
Nuphar Sm./Nymphaea L. ssp. — Water-Lily
water lyly(C21)

Menradix [meradix]
'anetum agreste'
?*Peucedanum ostruthium* (L.) Koch ? Master-Wort
anglice bissopeswort(B2) anglice bishopswort(B7)
anglice buschoppswort(C6) anglice byschoppeswort(C12)

Menta
Mentha L. ssp. Mint
gallice mynte(A1) mente, minte(A3) gallice mente(A4)
(m. alba) anglice horsment(B1) gallice, anglice mynte(B2) mente(B3)
mynte(B4) (m. aquatica) horsment(B3)
(m. domestica/ortulana alba) wyt mynt (domestica/ortulana rubea) rede
mynt(C1) (aquatica) horsmynt, spryngwort(C21) (domestica) anglice
tame mynt(C26) mynt, wodemynt(C30)

Menta Anana [m. amana]
Mentha aquatica L. 'Water Mint'
fisshment(C21) fisshment(C24) fyschment(C27)

Menta Romana
Mentha L. ssp. Mint
anglice wylde mynte or mynt Rome(C41)
Lit. Agnus Castus p.180, 22ff.

Menta Saracenica [m. cerasenica]
Chrysanthemum balsamita L. Costmary
costmarine(C2) anglice costmarine(C12) (et romana) whiȝt mynt(C21)

Mentastrum
Mentha sylvestris L. Horsemint
Mentha aquatica L. 'Water Mint'
gallice mentastre, anglice wyth mynte(A1) mentastre, wyte mynte(A3)
horseminte(A7) orsement(A10)
anglice brokmente vel horsement(B1) gallice mentastre, anglice wite
minte(B2) ho[r]ssemynte(B5) mentastyr, horsmynte(B8)
anglice watermynte(C3) gallice mentastre, anglice qwyte mynte(C18)
anglice white mynte(C22) anglice rede mynt or horsmynt(C26)
felde-mynte(C28)

Mentula Episcopi
Ranunculus ficaria L. Lesser Celandine, Pilewort
medewort(A3)
anglice medeflour(B2, B7) secundum aliquos melicia vel meliaxa, gallice

melise, anglice medewort(B9)
anglice medeflowre(C6) .ii. maner de mentule, anglice medflour(C18) water-folefote or medeflour(C21) moderwort(C22) medeflour þat growit in waters wit a brode lef and a ȝelow flour(C24) medeflour(C24) medeflour and hatz a ȝelw flour and a brod[. . .] and groweth in water(C27)
Lit. Alph., p.115, 29

Mentula Sacerdotis
Cichorium intybus L. — Chicory, Wild Succory
anglice sicorii(B2)
sircoure(C10) sirtour(C11) socorii(C16) anglice sicorii(C18) circoure(C21) sicoru(C23)

Mercurea
?*Mercurialis perennis* L./ *annua* L. — ? Dog's Mercury/ 'Annual Mercury'
anglice cme[r]worht(C15)

Mercurialis [mercuria]
Mercurialis perennis L./ *annua* L. — Dog's Mercury/ 'Annual Mercury'
mercurial, smerewort(A3) kerlekes*(A11)
millefoyle*, nosebledel*(B1) mercurial, anglice smerewort(B2) mercurial(B3) anglice smerwort(B7) gallice mercury, anglice smerdok(B9)
(maior) more mercurie, papwort (minor) mercurie(C1) anglice smeredoke(C3) mercuri, papwort, more smerwort(C21)
Lit. André I, p. 207 & II, p.159

Mespila [mespilia, mespulum, mespiles, mespilus, mespilum]
Mespilus germanica L. — Medlar
nepheles, openers(A3) gallice meles(A5)
anglice oponhers, medler(B2) gallice blabes, anglice medles(B7)
gallice blabeus, anglice medles(C5) gallice blaber, anglice medelez(C6) nephes, blabes, medles(C8) noeffles, melees(C10) anglice medlers(C12) gallice nefles, anglice medlers(C20) medles, anglice opyners(C22) anglice opyn-arses or medlers(C26)
Lit. B III, 179

Miacana
?*Anthriscus sylvestris* (L.) Hoffm. — ?Cow Parsley, Keck
woodphille(C21)

Micarea Solis
Calendula officinalis L. Pot Marigold
Cichorium intybus L. Chicory, Wild Succory
anglice rooden(C10)

Miconium [muconium, mico]
Papaver somniferum L. White Poppy
anglice wyth popy(A1) wyt popi(A3)
anglice wit popy(B2)
whit popy(C10) whyt popy(C26) whyte papye(C27)

Milium
Panicum miliaceum L. Millet
gallice semence(B2)
(medium) myle (minus) lasse myle(C1)

Milium Solis
Lithospermum officinale L. Gromwell
gromoyl(A3)
gallice et anglice gromoile(B2) gromyl(B3) gremele(B4) gallice et anglice gromille(B9)
gromyle(C22)

Millefolium
Achillea millefolium L. Yarrow, Milfoil
gallice milfoyl, anglice yarewe(A1) milfoil, yarowe(A3) gallice millefoyle, anglice yarowe(A4) anglice noseblede vel yarue(A6) anglice nasegras(A8) anglice nosebledeles(A9) anglice nosebledels(A12)
gallice milfoil, anglie ʒarue vel noseblede(B2) milfoil, nosebledels(B3) (m. minus) ambrosie(B3) neseblede(B4)
nosblede(C1) gallice mylfoyle, anglice yarw vel nosegrase(C6) mylfoile, yargh(C21) nosebledels(C23) yarowe or nosegres(C29) zarwe vel nesegras(C36) milfoyle, crebyld [sic](C38) nosebledde vel zarowe(C43)

Mira Solis
Calendula officinalis L. Pot Marigold
marigoolde(C21)

Mirefilon [marefilon]
Achillea millefolium L. Yarrow, Milfoil
noseblede(C1) nesebledell(C21)
Lit. B III, 105; André I, p. 214 (myriophyllon) & II, p.167

Mirica [merica]
Sarothamnus scoparius (L.) Wimmer ex Koch — Broom
Calluna vulgaris L. — Ling Heather
gallice genesta, anglice brome(A1) geneste, brom(A3)
anglice brom(B1) gallice geneste, anglice brom(B2) brom(B3)
brome(B4) brom(B5)
anglice brome(C3) lyng or bruer(C21)

Mirra
Commiphora myrrha Holmes — Myrrh
mirre(A3) gallice mirre(A4)
gallice et anglice mirre(B2)

Mirrea
Myrica gale L. — Sweet Gale, Bog Myrtle
mirte, goule(B2, C16)
goule(C23)

Mirta [mirtus]
Myrica gale L. — Sweet Gale, Bog Myrtle
Myrtus communis L. — Myrtleberry
Vaccinium myrtillus L. — Bilberry, Blaeberry, Whortleberry
gallice mirte, anglice gouuel vel goul vel horsmynte* vel wylde mynte*(A1) mirte, gowel vel goul, horsmynte*(A3) gowel(A5) anglice gaule(A6) anglice gawle(A8)
gallice gauge, anglice gawele(B1) anglice gouul(B2) anglice gaule(B7)
gowle(C2) tameryndys*(C8) gallice geuele(C8) mirtus est fructus þat ys clyped mirtillus, anglice hortene; mirta est fructus þat growyþ on mores .i. gaweil(C10) hurtene(C11) anglice cauel(C12) gayle(C16) gaule(C21) cuius semen vel fructus dicitur murtillis, blobeus(C22) mirete, gowel(C22) fructus dicitur mirtillus, blabeus, frehans(C23) it growet in more .i. gawl(C27) gawle(C27) gakel [corr. gawel](C32) hortilberies(C37) knotwort* oþer gasar*(C42)
Lit. B III, 141

Mirtillus [mirtillum, mirtilla]
Vaccinium myrtillus L. — Bilberry, Blaeberry, Whortleberry
Myrica gale L. — Sweet Gale, Bog Myrtle
gallice blabeaus, anglice frechanes(A1) blebeus, frochanes(A3) gallice

mirtiles, anglice frachans(A4) gauyl vel gayl(A5)
anglice howrte or gowl(C26)

Molena
Verbascum thapsus L. — Mullein
moleine(B3) softe(B5)
molen(C3) moleyne, sorel*(C10)
Lit. Macer pp. 222f.

Monica [monacha]
?*Rosmarinus officinalis* L. — ? Rosemary
?*Chrysanthemum segetum* L. — ? Corn Marigold
boyiun(A3)
budel(B2) budent[sic](B7)
budel(C16, C18, C23)
Lit. B II, 15

Monoclosios [monoclesios]
Artemisia vulgaris L. — Mugwort
anglice mugwort(C16, C18)
Lit. André I, p. 42 (Artemisia monoclonos) & II, p. 26

Montana [montanum]
Mentha pulegium L. — Pennyroyal
anglice broþerwort(A2, A8, A9) anglice bruþirwort(A6) anglice broþerworth(A12)
brotherwort(B3)
brotherwort(C2) sperewort*(C34) sperworte*(C37)
Lit. B III, 40f.

Montanum Minor
?*Aristolochia* L. ssp. — ? Birthwort
anglice smereworte(C3)

Mora
Rubus fruticosus agg. — Blackberry
?*Atropa bella-donna* L. — ? Deadly Nightshade
anglice molberien(C12) blakberies(C21) (m. celsi) mulbery(C21) (m. celsa) þe grete nichsode(C28) beth þe frutt of cikamore*, morus is þe tre(C30)

Mora Nigra
Rubus fruticosus agg./*caesius* L. — Blackberry/Dewberry

mores, blakeberie(A3)
blakberies or dewberies(C10) mores, blakeberye(C22)

Morella [merella]
Solanum nigrum L. 'Black Nightshade'
anglice houndeberie(A12)
morelle, mixplont, atterloth(B8) hunflours(B9)
morell et nightscode(C7) anglice atterlove(C20) morel, houndbery(C21) hundesbere(C22)

Morella Maior
Atropa bella-donna L. Deadly Nightshade, 'Great Morel' (m. maior and minor often confused)
anglice hundesbery(A1) hundesberie(A3) anglice hundesberiis(A8) anglice hundeberi(A9)
houndesberie(B2)
anglice morell(C3) anglice grete moral or dwale(C26)

Morella Minor
Solanum nigrum L. 'Black Nightshade', 'Petty Morel'
gallice morel, anglice attirlobery(A1) morele, attirloveberie(A3) (-minor)
gallice morele, anglice attirlovebery(A4)
gallice morele, anglice attuluberie(B2)
attuleþberynn vel honydesberynn(C6)

Morginella
?
gallice, anglice pympernell(C6)

Morsus Demonis [m. demonum]
Succisa pratensis Moench Devil's-Bit Scabious
herbe coppe(A3)
.i. maner de recope, gallice et anglice herbive(B2)
herbe coupe(C10) herbe recope(m)(C11) herbe coupe(C22) herbe ive(C23)

Morsus Diaboli
Succisa pratensis Moench Devil's-Bit Scabious
Leontodon autumnalis L. 'Autumnal Hawkbit'
anglice reycoype(A6) gallice recope(A8) gallice reucoupe(A9)
poukesbite(B3) gallice recope(B9)

pokebyte(C1) gallice recoupe, anglice bytingras(C6) recope(C7) reucopeu, bytongras(C8) un manere de recoupe, gallice et anglice herbayv(C16) un manere de recope, gallice et anglice erbyve(C18) devyll-bite or more herb ive(C21) anglice bugle*(C28) ffendis-bytt, reucoupe(C33) develes-bete, herbe ive(C35) gallice recoupe, anglice bitengras(C36) gallice recuper, herbeþat is lyke lyonnestothe(C37) poukebyte(C38) forbitte or more erbyve(C41) more ybitt[sic](C42)

Morsus Galline
Stellaria media (L.) Vill. — Chickweed
Anagallis arvensis L. — Scarlet Pimpernel
morone, chikenemete(A3)
chekynmete(B4)
anglice schekmete(C12) chekunwede(C21) morone, anglice chykemete(C22)

Mortagon
Lilium martagon L. — 'Martagon Lily'
valeriane*(C7)

Mortua
Solanum nigrum L. — 'Black Nightshade'
morell(C28)

Morus
Rubus fruticosus agg. — Blackberry
mulbery-tre(C6) molberye-tre, (fructus) molberye(C8)

Mubab
Chelidonium majus L. — Great Celandine
celidony, tetirwort(C21)

Muda [for herba munda?]
?
spurge, anglice sperewort(C16)
Lit. B III, 112f.

Muncaria [for matricaria?]
?*Artemisia vulgaris* L. — ? Mugwort
mugwode(C28)

Muscata
Asperula odorata L. — Sweet Woodruff

Erodium moschatum (L.) L'Hérit. 'Musk Storksbill', 'Herb Robert'
Geranium L. ssp. esp. *G. robertianum* L. Herb Robert
anglice wodrove(C6) woderove(C8) (maior) anglice herbe Robert(C16)
woderove(C21)

Muscus De Campo
?*Cyperus longus* L. ? Galingale
?*Asperula odorata* L. ? Sweet Woodruff
horstoth(C21)
see **Dens Equi**

Napei [naper, nappei, napium]
Sinapis L. Mustard
gallice la semence de mustarde(A1) la semence de mustarde(A3)
gallice la semence de mustard(B2)
mustard-seede(C10) gallice le semense de mustarde, anglice
scherlok-sede(C16) gallice la semence de mustard, anglice
cherloke-sed(C18) mustard-seed(C21)
Lit. André I, p. 216 (napy) & II, p.169

Napus [napis]
Brassica napus L. Rape
ney, nep(A2)
gallice nephele, anglice nepe(B1) napes(B4)
Lit. André I, p. 216 & II, p.169

Narcissus
?
houndistong*(C21)

Nardostachium [nardostochium, nardostancium]
Nardostachys jatamansi L. Spikenard
?*Cyperus longus* L. Galingale
spicanarde(A3)
spicanard(B2) spikirnade(B7) anglice spykenard(C6) spikanarde(C10)
spikenard(C21) anglice spykenard, englysche galyngalle(C26)
Lit. André I, pp. 216f & II, p.170

Naris Appolli
Mandragora officinarum L. Mandrake
mandrake(C21)
Lit. André I, p. 36 (Apollinaris) & II, p. 21

Nasturcium [narstucium, nastrucium, nastureticus]
Lepidium sativum L. Garden Cress
kersun(A11)
cresses(B3) tuncars(B8)
(n. album, domesticum, maius) towncarse(C1) aliud genus nasturcii quod vocatur frensch carsen vel carsen de boleyn(C6) crassen de huloyne[sic](C5) carsis of boleyne, modirwort*(C31) genus nasturcii est anglice french cresses vel cresses de boleyn(C36)

Nasturcium Agreste
Lepidium sativum L./ *campestre* (L.) R. Br. Garden Cress/Pepperwort
carsun savage, wylde carse(A3) kersunes(A4)
feldkerse(C2) anglice swynuscarsen .i. wyld carsen(C6) anglice feldecerse(C12) kersyn, wilde kerse(C22)

Nasturcium Aquaticum [narstucium a.]
Rorippa nasturtium-aquaticum L. Hayek Watercress
gallice kersune de funtayne(A1) watercarse(A3) kersunes(A4) kersouns(A6) gallice carsouns(A8) gallice cressun(A9) anglice cresson(A12)
anglice watercresses(B1) gallice kersoun de fountaigne, anglice waterkersoun(B2)
watercarse(C1) gallice cres de fonteyn(C16) (album) whi3t watercresse(C21) wel-crissis(C35)

Nasturcium gallicum [n. gallicanum]
Barbarea vulgaris R. Br. 'Winter Cress'
frenche carse(C1) frenche cresse, womans modrewort(C21) frensch kerse, women moderwort(C32) frensche cresse or woman medeworte(C41)

Nasturcium Mureligi
Nepeta cataria L. Cat-Mint
catmy[n]te, nept(C1) nept(C38)

Nasturcium Ortolanum [n. ortense]
Lepidium sativum L. Garden Cress
gallice kersune de ortil(A1) tuncarse(A3)
gallice cresses de curtil vel cardomon(B1) gallice kersoun de curtil, anglice tounkersen(B2)
tuncrissis(C35)

Nasturcium Porcinum
Capsella bursa-pastoris (L.) Medic. Shepherd's Purse
swynescressen(B1) sanguinarye(B1) (n.p. agreste) anglice
swineskerse(B2)
swynescarsis(C7) shepherdespurs(C10) sheperdyspowche vel
pikpurse(C11) stanche, anglice swynisgrasse(C16) stanche, anglice
[s]uinescarse(C18) sheperdspurs(C21) anglice herbe ive* or more
swynesgrasse(C26)

Nasturcium Pratinum
Lepidium heterophyllum Benth. 'Smith's Cress'
medicarse(C1) medowcresse(C21) medekerse(C32) medecresse(C41)

Naturella [for maturella]
Valeriana officinalis L. Valerian
hertwort, valeryan, lovewort(C21)

Nenufar [nenifar, nunfar]
Nymphaea alba L. White Water-Lily
Nuphar lutea (L.) Sm. Yellow Water-Lily
anglice waterlilye(A1) waterlilie(A4)
anglice waterlilie(B2)
(n. album) watterams (n. croceum) crocen(C1) anglice waterlylye(C6)
gallice lelie de ewe, anglice watirlelie(C16)
Lit. N195

Nepeta [nepita, nepta, nepten]
Nepeta cataria L. Cat-Mint
Lamium L. ssp. Dead-Nettle
katysme[n]te, nepte(A3) wylde netel(A5) anglice blinde nettel(A5)
cutemynte(A6) tonteme[n]te(A8)
sedmente(B1) nepte .i. cattesminte(B2) nepte(B3) calamente(B5) anglice
coumet [error?](B9)
katmynte(C1) catmynte(C1) gallice nept vel calamynte, anglice
cattisgras(C6) gallice nepte, calamynte, anglice cattesjoye(C8)
counteminte(C15) (minor) calamynte or cattesmynte(C27) tattesgres
[corr. cattesgres](C36)
Lit. André I, p. 218 & II, p.171

Nervene
Centaurea scabiosa L. 'Greater Knapweed'
Centaurea nigra L. 'Lesser Knapweed'
knapwort(C8)

Nigella
Nigella sativa L. Black Cumin
Agrostemma githago L. Corn Cockle
Lolium temulentum L. Darnel
nel, kokkel(A3) gallice neel, anglice cokel(A4) anglice cokel, gallice neel(A5)
gallice neel, anglice cockel(B1) gallice neele, anglice cockel(B2) cokkel(B3) popill, nele, cockyl(B4)
kockyl(C1) cokell, popell(C7) cokyl, popil, neel, wild sanegre(C21)

Nigra Spina
Prunus spinosa L. Blackthorn
anglice blacthorn(B1)

Nimphea
Nymphaea alba L. White Water-Lily
Nuphar lutea (L.) Sm. Yellow Water-Lily
?*Saponaria officinalis* L. ? Soapwort
anglice edok(B1) crousope secundum quosdam(B3)
heddoc, rote scabious(C28)
Lit. Alph., p.125, 28

Nimphea Aquatica
Saponaria officinalis L. Soapwort
crousop(B1)
crousope, fomedok(C7)
Lit. Alph., p.125, 28

Nitrago [vitrago]
Parietaria diffusa Mert. & Koch Pellitory-of-the-Wall
paritarie(B3)
anglice peretorye(C12)
Lit. Alph., p.125, 14

Nitrucia
Centaurea nigra L. 'Lesser Knapweed'
iacea nigra vel rubia minor, romani vocant matfeloun, anglici cappewort(C43)

Nosubullus
?*Glechoma hederacea* L. ? Ground Ivy
anglice heyhove(C16)

Nux [simplex, simpliciter]
Juglans regia L. Walnut
noyz de noge, walnothe(A3)
gallice nugage, anglice walnote(B2)
anglice walnote(C8) (maior) walnote(C21) noiz de noge, walnote(C22)

Nux Avellana
Corylus avellana L. Hazelnut
noyce de caudre, haysell(C7)

Nux Coruli
Corylus avellana L. Hazelnut
gallice noyz de coudre(B2)
id est quod avelana, gallice noy de codre, anglice smal nottis(C16)

Nux Gallica
Juglans regia L. Walnut
anglice walnotys or walysh notys(C26)
Lit. André II, p.173

Nux Indica
Cocos nucifera L. Coconut
noyz de Inde(A3) gallice noiz de Ynde(A4)
gallice noiz de Ynde(B2)
Lit. N196

Nux Longa
Prunus amygdalus Batsch Almond
almandes(C10) almundis(C11)

Nux Miristica
Myristica fragrans Houtt. Nutmeg
noyz mugete(A3)
gallice noiz de muge(B2)
notemuges(C21)

Nux Muscata
Myristica fragrans Houtt. Nutmeg
gallice noyz de myge(A1) noyz mugete(A3) gallice noiz mugette(A4)
notemuge(C8) notemug(C21) notmwge(C41)

Nux Pontica
Corylus avellana L. Hazelnut

gallice noyz de coudre(A1) noyz de coudre(A3)
gallice noiz de coudre(B2)
notes of hasel tree(C10) anglice walnot*(C16) hasell(C21) noyz de coudre(C22)

Nux Vomica
Castanea sativa Mill. — Chestnut
Strychnos nux-vomica L. — 'Poison Nut' ('castanea indica')
chasteynes(C10) chesteynys(C11) chesteynes(C21) chastyn(C35)
Lit. N196f.

Obstrutio
?*Saponaria officinalis* L. — ? Soapwort
?*Betonica officinalis* L. — ? Betony
?*Peucedanum ostruthium* (L.) Koch — ? Master-Wort
anglice biscopiswort(A2) bissopishisworte[sic](A3)
Lit. André I, p. 45 (astrutium) & II, p. 29; B III, 29

Obtalmicum
?*Calamintha ascendens* Jord. — ? 'Common Calamint'
calamynt(C21)
Lit. André II, p. 210 (ptarmice); Alph., p.127, 8

Occiminium [ozimum]
Urtica L. ssp. — Nettle
urtie, nettle(A3)
a seed [corr. reed?] netel(C10)
Lit. André I, p. 224 (ocimum) & II, p.175; B III, 172f.; Alph., p.133, 40 (ozminim)

Ocimum [occimum, ozimum, osmum, osinum]
Viscum album L. — Mistletoe
dragaunz*, naddirwort* vel lambesfot*(B1) mistelto(B3)
zetuwale*(C2) misteles(C12) mystell(C21) mistulto(C21) mistell(C28) nedirwort*(C28)
Lit. B III, 172 & 175

Oculus Bovis
Chrysanthemum leucanthemum L. — Ox-Eye Daisy
ruþerestonge*(B5)
oxe eye(C1) .i. bothon(C10) ox-eye(C21) bothun(C27) oxe yȝe(C37)
oxis eye(C41)

Oculus Christi
Calendula officinalis L. Pot Marigold
Delphinium ambiguum L. Larkspur
?*Chrysanthemum leucanthemum* L. ? Ox-Eye Daisy
?*Salvia sclarea* L./ *verbenaca* L. ? Clary
anglice seint Mari rode vel Cristis-ey(A1) solsekyl(A3) anglice laverockespore(A6) laverokespure(A9) anglice clavrokespore(A12) gallice solsicle, anglice seynte Marie rode(B2) oyle[sic] de criste-eye(B8) anglice bisseplese(B9)
anglice laverokespere(C3) anglice schilwort(C5) anglice scalleworth(C6) anglice oxeseye(C12) gallice solsecle, anglice redwort(C16) okyl Christi(C21) mersewort, claverokspor(C34) schallegres(C36) mersewort anglice claveroksporre(C37) Cristes yee(C37) scabworte(C39) wilde worte oþer scabwort(C42)

Oculus Consulis [o. consulus]
?*Chrysanthemum leucanthemum* L. ? Ox-Eye Daisy
?*Salvia sclarea* L./ *verbenaca* L. ? Clary
oyle de beuf, cowslype(C7) clarrey(C21) horsmynte*(C28) gallice cicuse[?](C39)
Lit. Alph., p.126, 35

Oculus Diei
Bellis perennis L. Daisy
daysy, consounde(C28)

Oculus Leonis
Bellis perennis L. Daisy
anglice daysye(C26)

Oculus Lucidus
Lonicera caprifolium L./ *periclymenum* L. 'Perfoliate Honeysuckle'/ Honeysuckle
trifoyl(A7)
anglice þe jus of þe grete wodbynd(C26)

Oculus Lucii
Lonicera periclymenum L./ *caprifolium* L. Honeysuckle/ 'Perfoliate Honeysuckle'
wodebonde(A3)
woodbynd(C21)

Oculus Populi
Populus L. ssp. Poplar-Bud
borjons of popler(C10) popeler(C16) the beriez of ye popler(C21)
burjuones of pepul(C24)

Oculus Sconsus
? ?
okelescons(B1)
Lit. in the inflated version of Macer in MS London, B.L. Royal 12 B XII
f256r there are verses 'Est oculo sconsa [corr. sconso] vis fervida
siccaque valde'

Oculus Solis
?*Reseda luteola* L. ? Dyer's Rocket, Weld
?*Salvia sclarea/verbenaca* L. ? Clary
anglice wolde(C28) centrum gali .i. cokkispore(C35)
Lit. cf. André II, p.176

Oculus Sponsi
Conium maculatum L. Hemlock, Herb Bennet
anglice hemlok or herbe benet(C26)

Oculus Vacce
?*Chrysanthemum leucanthemum* L. ? Ox-Eye Daisy
cow eye(C1) cow ye(C38)

Ointerele [ointerell, ointere, ourtele, ourte, oniterrele]
Prunella vulgaris L. & var. applied Self-Heal
anglice selfhele(B2) anglice selfheile(B7)
anglice selfhele(C6) anglice selfeue [corr. selfele](C12) anglice
selfhele(C16) anglice selfhele(C18) .i. selhele(C23)
Lit. Alph., p.193 (unctuosa)

Oleum Olive
Oil of Oliva europaea L. Olive Oil
gallice oyle de olive(A1) oile de olive(A3)
gallice oyle de olive(B2)

Olibanum [libanum]
Boswellia thurifera L. Frankincense
fraunc encense(A1) frank encens, stor(A3) franc encens, stor(A3) gallice
fraunk encens(A4) franc ensens(A7)
gallice franc encens, anglice ȝwit stor(B2) gallice franc ensenz, anglice

store(B2) frankencense(B4)
franc encens(C2) gallice frankencens, anglice whit stor(C16)

Olus
Brassica oleracea l. Wild Cabbage
wort(C21)

Olusatrum [olisatrum, olixatrum, olisarum, olistirum, olasatrum]
Smyrnium olusatrum L. Alexanders, 'Horse-Parsley'
persil(A3)
gallice alisaundre, anglice stanmarch(B2) anglice alisaundre(B7)
anglice alysaunder vel stanmerch(C6) stanmerch(C8) alisaundre, s[t]anmarche(C21)

Omnimorbia
?*Teucrium polium* L. ? Mediterranean Germander
polipodie*(B3)
Lit. André I, p. 227 & II, p.177

Opiteos
Pteridium aquilinum (L.) Kuhn & other ferns
the comon brake(C21)

Opium
Papaver somniferum L. Juice of Opium Poppy
gallice jus de popy(A1) jus de popy(A3)
gallice le jus de popie(B2)
dweledrenk(C7) (thebaicum) þe juse of whit popy(C21)

Opocissi
Hedera helix L. Ivy-Gum
gume de ere(A3)
gallice gumme de ere, anglice yvi(B2)

Opoponax
[*Opoponax* L. Opoponax]
ffaytourswort*(C30)

Orbicularis [orbicularius, orbiculare]
Cyclamen hederifolium Ait. Sowbread
?*Arctium lappa* L. ? Burdock
clete*(A6) clote*(A9) anglice clote*(A12)
clote*(B3)

anglice clete*(C3) cli[th]e*(C7) anglice clote*(C12) erthnote, dyllenote, halswort(C21)
Lit. B III, 209 (slite) & II, 105f; André I, p. 229 & II, p.180

Ordeum
Hordeum L. ssp. Barley
orge, balrich(A3)
anglice barlik(B1) gallice orge, anglice barlich(B2)
barlicle(C1)

Oriatanum
Inula helenium L. Elecampane
elle(B1)
Lit. B III, 82f.

Origanum [origonum]
Origanum vulgare L. Marjoram
Mentha pulegium L. Pennyroyal
Thymus serpyllum L. Wild Thyme
gallice origane, anglice puliol reale(A1) puliol real, origan(A3) gallice puliol reale(A4) gallice puliol real(A5) anglice hulwort(A6)
gallice puliole real(B2) organe(B3) orgoyne(B4) anglice pulliol real(B5)
anglice organ(C1) brokwort*(C2) gallice puliol real vel graunt puliol, anglice grete boterwort[sic](C5) puliol real, grant puliole, gret broyswort*(C18) anglice brokwort*(C12) gallice grant pulele, anglice þe gret broþ[er]wort(C16) gallice la grant puliole, anglice þe grete brotherwort(C18) puleal reall, cyrchewort(C21) le grant puliol, ye grete broþerwort(C23) puliol real, wodemynt(C24) orgone(C27) puriol reall(C28) orgon(C31) gretelwort(C43)

Origeron [origenen]
?*Senecio vulgaris* L. ? Groundsel
senechion, groundeswilye(C8) anglice groundswilie(C19)

Orminum
Urtica L. ssp. Nettle
gallice urtie, .iii. maner de urtie, anglice nettle(B2)
gallice urti(c)e, anglice netell, .iii. maner de urtise(C16)
Lit. cf. André I, p.165 (horminum) & II, p.126; see Alph., p.131, 30; N198

Ornus
Carpinus betulus L. Hornbeam
horn(C8) holin tre*(C12)

Orobus [erbus, herbus]
Lathyrus L. & *Vicia* L. ssp. Vetch, Tare
anglice fiches(A2) veches vel vechis(A3) anglice mispese(A6) anglice musepose(A8) anglice mousepese(A9) anglice mousepeyse(A12) mosepese(B3) mousepese(B3) anglice muspese(B9) (maior, purpureus) haþere (minor, albus) mowsepuse(C1) muspipe(C3) anglice mousefychches(C6) mousfecche(C8) mospece(C10) mousepese(C11) anglice moushere*(C12) gallice peluet*, anglice mousere*(C12) mosepose(C15) gallice vesses(C20) mouspese(C21) (de Hispania) mouspese of Spain(C21) vechies(C22) vesce, muspese(C23) anglice tame wacches(C26) mousefych(C29) mosepese, tilles(C31) mussepype(C34) wild fechis(C35) moussepipe(C37)

Orton
?*Verbena officinalis* L. ? Vervain
verveyne(A3)
gallice verveyne(B2)
anglice verveyne(C10, C22)

Os Dactuli
Phoenix dactylifera L. Fruit of the Date Palm
pire de date(B4)
Lit. André I, p. 231

Osa
Alkanna tinctoria (L.) Tausch Alkanet
Pentaglottis sempervirens (L.) Tausch Alkanet
rede alknyt(C21)

Osmites
Mentha L. ssp. Mint
quite mynt(C21)
Lit. André I, p. 232 & II, p.183

Osmius
?*Geum urbanum* L. ? Wood Avens
avence or harefote(C21)

Osmunda [osimunda, osamunda]
Osmunda regalis L. — Royal Fern
Polypodium vulgare L. — 'Polypody'
osmunde, eververne vel evyrverne(A3)
gallice omunde, anglice eververn(B2) omounde(B3) omonde(B4)
gallice osmounde, anglice everfern vel mulbery-tre*(C5) osmunde .i. ysenerd(C7) omo[n]de, evyrfern, multerdene(C8) polipodie ys fern of an oke(C10) osmund, wodebrake, polipodi(C21) rayge, wild brake(C21) osemounde, anglice eververne(C22) overferne on þe ook, pollipodye(C27) wild brake(C35) overfern vel milterdene[?](C36) mistilte* oþer arbuste*(C42) osmunde oþer diche-verne(C42)

Ostriago [ostrago, ostraga]
?*Stellaria holostea* L. — ? 'Greater Stitchwort'
?*Sambucus ebulus* L. — ? Danewort
anglice stikwurt(A11)
anglice lichewort vel liverwort*(B1) walwort(B3)
anglice walwort(C12) pigla, stechewort(C16) lithewort(C28) anglice smallache wort* [corr. small wort?] and pigill(C29)
Lit. B III, 218 & 163

Ostriagum
?*Plantago coronopus* L. — ? Buck's-Horn Plantain
ive, oþer herbe yve oþer hertishorne(C42)
Lit. Wright & Wülcker, A.S. & O.E. Vocabularies I, p. 558, 21 ('ostragium .i. herbyve .i. liþewurt')

Ostribrum [ostribium]
?*Sinapis arvensis* L. — ? Charlock
?*Raphanus raphanistrum* L. — ? Wild Radish, White Charlock
anglice cherlok(C4) cherlok(C7)

Ostrutium [ostricium, ostrecium]
?*Betonica officinalis* L. — ? Betony
?*Peucedanum ostruthium* (L.) Koch — ? Master-Wort
anglice biscopworth(A12)
brok-karse(B5)
bisshipiswort(C7) beteyn, bisshopiswort(C21)
Lit. B III, 29. See **Strutium**

Osturce [osture, ostrusse, osturse]
Brassica oleracea L. — Wild Cabbage

wilde cole(C5) anglice wyld coule(C6) wilde coul(C8) wilde caule(C23)
Lit. see **Strutium**

Othonia
Chelidonium majus L. Greater Celandine
celidony, tetirwort(C21)
Lit. André I, p. 232 (othone) & II, p.183

Oxicantum
Berberis vulgaris L. Barberry
anglice barberyns(B7)
anglice berberyes(C5) berberius(C6) verberys(C10) barberes(C21) berberes(C24)
Lit. cf. André I, p. 233 & II, p.183; N199

Oxifenicia [oxifinicia]
Tamarindus indica L. Tamarind
gallice tamarindis(A4)
sorel*(C10) anglice the berys of quyckbeme(C26) tamaryndes(C27)
Lit. N197

Oxilappatium [occilapacium]
Rumex acetosa L./ *acetosella* L./*obtusifolius* L. Sorrel/Sheep's Sorrel/ 'Broad-Leaved Dock'
surel, surdoke(A3) docke(A11)
ovenscharp(B1) gallice eysel vel sorel, anglice sourdoc(B2)
surell(C7) gallice exsel vel surele, anglice souredock(C18) oversarp [corr. ovensarpe]-dok(C28)

Oxilonon [for oxigonon]
Papaver somniferum L. Opium Poppy
whit chesbolle(C21)
Lit. André I, p. 233 (oxygonon)

Oxygia
Urtica L. ssp. Nettle
þe sede of nettyll(C27)

Ozireos
Antirrhinum orontium L./*majus* L. Weasel's Snout/Snapdragon
houndisheed(C21)
Lit. André I, p. 232 (osireos tafe) & II, p.182

Pacientia
?*Rumex patientia* L. ? 'Patience Dock'
paciens(C4, C21)

Palatium Leporis
Geum urbanum L. Wood Avens
anglice avence(C26)

Paliurus [palurus]
Cirsium Mill. ssp. Thistle
Delphinium consolida L. 'Forking Larkspur'
calketrappe(B1)
herba spinosa, anglice podyll(C28)
Lit. cf. André I, p. 234 & II, p.185

Palla Marina [pella m.]
?*Pulmonaria officinalis* L. ? Lung-Wort
bellewort(C8) bellwed(C10) anglice longwort(C26)
Lit. Alph., p.137, 7 ('palla marina arbor est')

Palma Christi
Ricinus communis L. 'Castor-Oil Plant'
Lithospermum officinale L. Gromwell
Artemisia vulgaris L. Mugwort
Leonurus cardiaca L. Motherwort
Lamium purpureum L. 'Red Dead-Nettle'
gallice grumyle(A1) anglice moderwort(A2) gromoil(A3) anglice mouderwort(A6) anglice moderwort(A8, A9, A12)
gallice gromyl, anglice ragged moderwort(B2) moderwort(B3) anglice mod[er]wort(B9)
mannys moderwort (silvatica, media) stedfast(us)(C1) mannys modrewort(C4) anglice moderwort, et secundum aliquos dicitur milium solis, gromyl; secundum aliquos ragid moderwort in prima significatione, vel mogwed(C5) rede docke*(C8) anglice gromelye secundum quosdam, sed melius credo quod palma Christi similis est archangele[sic] . . . anglice mogwort(C12) gallice paume de Crist(C12) anglice raggit mode[r]wort(C16) anglice raggede moderwert[sic](C18) paume dowe [= d'ome?](C21) gromyle(C22) anglice ragwort or mannys moderwort: this herbe is lyke to archangel but the levys ben brodder and more thyk as .v. fingers with a .iiii. square stalke and somwhat blacke(C26) similis est archangelice sed folia habet maiora multo plus spissa, stipitem habet quadratum aliquantulum .i. anglice mogewort(C37) anglice crystispalme,

mannys-moderworte or stidfaste(C41)
Lit. B III, 173; Agnus Castus p. 240

Pampinus
Vitis vinifera L. Vine-Leaf
foille de vingne(A3)
gallice foyle de vine(B2)
vynelef(C8) ye les [corr. lef of] vyne(C21)

Panis Cuculi
Oxalis acetosella L. Wood Sorrel, 'Cuckoo's-Meat'
anglice wodesoure(A1)
anglice wodesbine*(B2) wodesoure(B3, B5) payn a cucuelė, wodesoure(B8)
cokkowbreed vel stubwort(C2) wodesowre(C4) cokesour, stobwort, sorel dé boys(C21) stabwort [sic](C34) scabbewort(C37)

Panis Porcinus [p. porci]
Cyclamen hederifolium Ait. Sowbread
erþenotes(C10) erthnote, dilnote, halsnote [corr. halswort?](C21) dilnote(C35)

Panphiligos [pamphilagas, pansiligos, pamfilagos, panfilago, pamphilegos, pamfiligos, panfilion, ponfilion]
Verbascum thapsus L. 'Great Mullein'
anglice feldworth(B1) gallice moleigne, anglice feltwort(B2) moleyne(B3) anglice feltwort(B7)
gallice moleyn, anglice feldworth(C6) molyn(C10) gallice moleyne, anglice fylwort(C16) lungwort, moleyn, ffeltwort(C21) moleine, feldwort(C22)
Lit. Alph., p.140, 2

Papaver
Papaver L. ssp. Poppy
gallice popy, anglice chessebole(A4) anglice popi(A6) anglice popie(A9, A12)
gallice pouncel, anglice popie(B1) anglice popie(B2) popie(B3) popi(B5) whyt popy, blak and red(B8) gallice [sic] popi, anglice chessebolle(B9)
(album, domesticum, maius) popy whyt (nigrum domesticum agreste) blac popy(C1) anglice chesbols(C3) (album) blank susement, whit popy, (nigrum) noir susement, blak popy(C10) chessebolle(C14) (rubeum,

domesticum) red chesbolle, popy (albus) quyght popy (rubeus, agrestis) wild chesbolle(C21)
Lit. N200

Pappa Muris
Sedum L. ssp. Stonecrop
stoncrop(B3)

Paranetum
Anethum graveolens L. Dill
dile(A2)

Paratella [paradilla, paradella]
Rumex hydrolapathum Huds. 'Great Water Dock'
Rumex sanguineus L. 'Red-Veined Dock'
parele, docke(B1) pareel, dokke(B8)
gallice parelle, dokke(C2) gallice et anglice dok vel red doke(C5) parel, dok(C7) gallice paratell, anglice dokke(C12) reed dokkis(C21)
Lit. André I, p. 238 (parada)

Pariciasos
Polygonum persicaria L. 'Persicaria'
Polygonum hydropiper L. Water-Pepper
persucaria minor, ersmert, culrage(C1)

Paritaria
[*Parietaria diffusa* Mert. & Koch Pellitory-of-the-Wall]
anglice mousere*(B2)
anglice mousere*, gallice peluetre*(C18) (maior) dropwort*(C21)

Paritoria [paritaria]
Parietaria diffusa Mert. & Koch Pellitory-of-the-Wall
?*Sedum acre* L. ? Wall-Pepper
paritorie(B1) gallice et anglice paritorie(B2) peritarie(B3) peritory(B4) anglice paritari(B9)
b[i]therwort*(C1) anglice paritory(C3) gallice et anglice parytorie(C6) paritorie, stonhoue [corr. stonhore?] walwort(C10) anglice peretorye vel walwort(C12) lychewort, hemerwort(C32) paritarie .i. stonhore(C35) parytorie vel lychewort(C38) paratory oþer lithwort other hem[er]wort(C42)

Parthenocidos [parimocidos, perumasidos, perimocidis, percinotodos]
Mercurialis annua L. 'Annual Mercury'

Mercurialis perennis L. Dog's Mercury
gallice mercurial, anglice smerewort(B2)
gallice mercurial, anglice smerwort(C16) mercuri, papwort(C21)
Lit. André I, p. 239 (parthenium) & II, p.189; Alph., p.138, 21

Passula [passulus]
Vitis vinifera L. Grape
resyns(A)
gallice reisynz(B2)
anglice reysonus(C6) reysenges(C21)

Pastinaca [pastina]
Pastinaca sativa L. Wild Parsnip
Daucus carota L. Wild Carrot
Heracleum sphondylium L. Cow Parsnip
Anthriscus sylvestris (L.) Hoffm. Cow Parsley
anglice skyrewyt(A6) anglice wilde chirwit(A8) anglice skirewit(A9) anglice scyrwit(A12)
wodechervelle vel schirwyth(B1) gallice pestnere, anglice wilde skirezwit(B2) wolemore(B3) chervele(B4)
pasnepe (agrestis) wyld pasnepe (pastina) skyrwhyth(C1) anglice pastnepps(C3) (domestica) passenepe (agrestis) wilde passenyp(C4) anglice pasturnepe(C6) chervile(C7) anglice wild scirwhit(C15) gallice pestnere, anglice skyrewhit(C18) pistenepe(C22) pasnep(C28) (agrestis) evelþroten or cow-wort(C31)
Lit. B III, 249 & 265f; N201

Pastinaca Cretica
Pastinaca sativa L. Wild Parsnip
?*Peucedanum officinale* L. ? Hog's Fennel
anglice pasturnate or houndisdil(C26)

Patersion
Filipendula vulgaris Moench Dropwort
filipendule, dropwort(C21)
Lit. Alph., p.135, 23

Patrucia
?
romani vocant eam patracion, anglici hisenerde(C43)

Pectes
Symphytum officinale L./ *tuberosum* L. Comfrey/'Tuberous Comfrey'

comfery(C21)
Lit. André I, p. 241 (pecte) & II, p.190

Pediciados
Parietaria diffusa Mert. & Koch Pellitory-of-the-Wall
perytory, hemerwort(C4)
Lit. André I, p. 242 (perdicias) & II, p.192; B III, 122. see **Perciados**

Pegma [?]
Fraxinus excelsior L. Ash
gallice freyne, anglice esh(B2)
gallice freine(C5) gallice freyne, anglice asshe(C16)

Pellitaica [?]
Polygonum hydropiper L. Water-Pepper
culerage(C22)

Pellitricus
Adiantum capillus-veneris L. Maidenhair-Fern
maydenher(C7)
Lit. André I, p. 257 (polytrichon) & II, p. 205 (polytrichum)

Pelotronia
Centaurium erythraea Rafn Common Centaury
more centory, erthgalle(C21)
Lit. André I, p. 255 (plectronia) & II, p. 202

Penidarum
?*Parietaria* L. spp. ? Pellitory
.i. maner de peletre(B2)

Pentadactulus
?*Potentilla reptans* L. ? 'Creeping Cinquefoil'
swynestale(C10) swynystayle(C11) swynescalle(C24) swynestailles(C27)
Lit. André, p. 241 (pentadactylon) & II, p.191; N201

Pentafilon [pentaphilus]
Potentilla reptans L. 'Creeping Cinquefoil'
gallice quintefoyl, anglice five-levede gras(A1) quintefoil(A3) gallice quintefoyl(A4)
quintefoyle, anglice fivelef(B1) gallice cincfoil, anglice fivelef(B2) fyflef(B3)
cinquefoile(C2) anglice syngfewle(C3) þe fyve-levyd grasse(C4) gallice

quintfoyl(C6) quinkefoil, vyf-leved gras(C8) quinkfoile(C21)
Lit. André I, p. 241 (pentaphyllon) & II, p.191

Pepilus
?*Apera spica-venti* (L.) Beauv. ? 'Silky Apera'
anglice wyndgresse(B9)

Pepones
Cucumis melo L. Melon
melones(C10)

Perciados
Parietaria diffusa Mert. & Koch Pellitory-of-the-Wall
?*Sedum acre* L. ? Wall-Pepper
paritori, stonehoue [corr. stonehore?](C21)
Lit. André I, p. 242 (perdicias) & II, p.192. see **Pediciados**

Perdicalis (Herba) [perdicularis, perdicialis]
Parietaria diffusa Mert. & Koch Pellitory-of-the-Wall
perdical, hamerwort(B1) ribwort*(B5)
(maior) peritorie(C1) ribwort*(C8) porrettory, litewort, hemerwort(C21)
anglice parytary; hit is an herbe with whiche glasse is clensyd(C26)
Lit. André I, p. 242 & II, 192; B III, 122

Perdiciados
[*Parietaria diffusa* Mert. & Koch Pellitory-of-the-Wall]
wodebynde*(C21)
Lit. André I, p. 242 (perdicias) & II, p.192

Periclymenon [perichimenon, parichemenon, parichimeon]
Lonicera periclymenum L. Honeysuckle
gallice cheverefoile, anglice wythewynde(A1)
gallice cheverfoil(B2)
wodebynde(C10) cheverfoyle(C22)

Periponon
?*Betonica officinalis* L. ? Betony
beteyn or bisshoppiswort(C21)

Peristerion [peristhemeron]
Verbena officinalis L. Vervain
anglice dowesbell(B1) cheverefoile*(B2)

culerage*(C10) verveyn(C21) vervena, fle(t)wurt(C23) culrage*(C24)
Lit. Alph., p. 43, 5; André I, p. 243 & II, p.193

Persa
Majorana hortensis Moench — Sweet Marjoram
Origanum vulgare L. — Marjoram
muchwort(C10) mocheworte(C11) majoran(C21) methwort(C24) muchwort(C27)
Lit. Alph., p.141, 33; N202

Persicaria [persicarium, percitoria, persucaria, perisicaria]
Polygonum persicaria L. — 'Persicaria', Willow-Weed
Polygonum hydropiper L. — Water-Pepper
gallice culerage(A2) gallice culrage, anglice hirssmerte(A6) gallice culrage(A8) gallice culerage, anglice waterwede vel hersmert(A9) gallice culrage, anglice waterwede vel red bne [corr. kne] vel ersmert(A12) culyerage(B1) gallice culrage, anglice dropewort(B2) culrage(B3) kulrage(B5) gallice culyrage, anglice wat[er]wede vel red bele [corr. kne] vel hersemerte(B9) (maior) ersmert, culrage, biterwort (minor, pratina) culrage(C1) (maior) dropwort(C2) culurage, red kne, waterwod(C3) bitterwort, culrage, ersmart(C4) anglice dropworth(C6) (minor) gallice culrage(C6) dropwort(C10) (minor) culrage(C11) (minor) gallice culrage, anglice arssmerte(C12) gallice calrage, waterwede(C16) (maior) anglice dropewort (minor) culrage, anglice erssmert(C18) culrage, arsmert, bitterwort(C21) droppewort(C27) hardocke, culrage and rap-coyle(B33) (maior) dropwort (minor) culrage vel waterwed(C36) culrage, arsmethe(C42)

Persicus
Prunus persica (L.) Batsch — Peach
gallice pecher(A5)
gallice persecher(C5) gallice pescher(C6) anglice pescher-tre(C12) peccher(C22)

Personacia [personantia]
var. appl. to large-leaved plants incl. burdock, beet, water-lily, darnel.
drox, anglice drauch(B1) betes(B3) drauk(B5)
elfrone, alii dicunt flotwort(C2) flokwort(C7) elfyone, alii dicunt flotwort(C12)
Lit. André I, p. 244 (personacia & personatum folium) & II, p.194; Alph., pp. 54, 13 & 68, 16; B III, 27 & 50f.

Pervenca [pervingca]
Vinca minor L. Lesser Periwinkle
pervynke(B4) pervinke(B8)
pervenk(C1) parvynk(C4)

Pes Alaude
?*Delphinium consolida* L. ? 'Forking Larkspur'
larkefote(C7)

Pes Ancipitris [p. accipitris]
Geranium L. ssp. esp. *G. columbinum* L. Cranesbill
?*Aquilegia vulgaris* L. ? Columbine
anglice hauekesfot(B2)
anglice rosgrasse . . . anglice haukesfote(C5) gallice [sic] rosegras, anglice haukysfote(C6) columbyne(C7) haukfote(C21) anglice herbe Robert(C26) rosegres(C36)

Pes Arietis
?*Ranunculus acris* L. ? 'Meadow Buttercup'
ramysfote, lodeworte(C42)
Lit. Agnus Castus, p.199, 8ff.

Pes Cervi
?*Conopodium majus* (Gouan) Loret ? Pignut, Earthnut
hertisfote(C21)
Lit. Alph., p. 40, 26

Pes Columbe [p. columbina]
Geranium L. ssp. esp.*G. molle* (Dove's-Foot) Cranesbill
?*Galium cruciata* (L.) Scop. ? Crosswort
?*Aquilegia vulgaris* L. ? Columbine
gallice pé de columbe, anglice culverfot(B2) kolverfot(B5) colverfote(C1) masewort(C2) herbe Robert(C5) spurge .i. maswort idem vel renewrt(C7) herbe Robeud, masewort(C8) anglice masewort(C12) anglice colverefot, sperge(C20) dowfote(C21) anglice herb Robert þe lasse, culverfot or dovefote or spurge(C26) may[s]worte(C34) mawort or columbyne(C37)

Pes Columbinus
?
herbe yve(A2)

Pes Cornicis
?*Ranunculus acris* L. ? 'Meadow Buttercup'
revenysfote(C1) ravensfote, ledewort [for lodewort](C21)

Pes Corvi
Ranunculus acris L./*aquatilis* L. 'Meadow Buttercup'/ 'Water Crowfoot'
?*Anemone nemorosa* L. ? Wood Anemone
gallice pé de corf, anglice crowefot(B2) ravenesfote(B4) revenisfot(B5) crakefot(C3) ravenesfote(C4) white crowfote(C4) ravenysfote, lodewort(C4) anglice crowfot(C6) ravensfote, remmesfote(C7) crowesfote(C10) gallice pee de corf, anglice crawfote(C16) (silvaticus) quite crowefote(C21) (agrestis) crowefote (aquaticus) water-crowfote(C21) crowisfote(C24)

Pes Corvi Silvatica/Silvaticus
?*Ranunculus* L. ssp. ? 'Crowfoot'
?*Sanicula europaea* L. ? Sanicle
wodcrowfot(C1) whyte crowfot(C32) wilde crowfote(C38)

Pes Galli
?*Aquilegia vulgaris* L. ? Columbine
?*Ranunculus acris* L. ? 'Meadow Buttercup'
columbine, cokfote(C21)

Pes Leonis [p. lionis]
Alchemilla vulgaris agg. Lady's Mantle
?*Leontice leontopetalum* L. ? 'Lion's Leaf'
gallice pé de lyoun, anglice lyounesfot vel souþistel*(B2) lyonesfote(B4) (minor) peedlyon(C1) pé de lyon(C4) gallice pee de lyon(C6) mathewort(C7) gallice pé de lyon, anglice soughethistel*(C18) maȝewort(C19) lionnesfote(C37)

Pes Leporis
Geum urbanum L. Wood Avens, Herb Bennet
anglice harefote(A1)
gallice avence, anglice harefot(B2) gallice pé de levere, anglice harefot(B2) avence(B3)
anglice harefoote(C6) harefot(C21)

Pes Mentularum
?
pé de coylun(A5)

Pes Nisi

?*Adiantum capillus-veneris* L.	? Maidenhair-Fern
?*Asplenium trichomanes* L.	? 'Maidenhair Spleenwort'
?*Asplenium adiantum-nigrum* L.	? 'Black Spleenwort'

sperhaukesfote(C10) sperowhaukefote(C11) hawkisfote(C24) sperhaukefote(C27) after sum men it is the kynde of a water kerse(C37)
Lit. cf. Alph., p.140, 32

Pes Pulli

Portulaca oleracea L./*sativa* L.	Purslane
Tussilago farfara L.	Coltsfoot
?*Nuphar lutea* (L.) Sm.	? Yellow Water-Lily
Ranunculus acris L.	'Meadow Buttercup'
?*Ranunculus ficaria* L.	? Lesser Celandine

gallice pé de poleyn, anglice coltisfot(A2)
gallice pé de poleyn(B2)
(agrestis maior) wolwfote (pratinus medius) ȝe[le]w cray, mareblowre(C1) (aquaticus) mentula episcopi, watur-folefote (praticum!) buldes, mareb[. . .], lobbe, yelow cray(C4) gallice pé de puleyn, anglice coltfote vel dunne [hof?](C5) gallice pee de peleyn[sic], anglice coltfote vel annye [corr. dunne?](C6) anglice purcelane(C12) (agreste) folefote (aquaticus) water-folefote(C32) (pratinus) medow-folesfote, yelwe crowe(C21) water-folefot(C32) (pratina) made-folefot, ȝelewe cray(C32) gallice pé de polayn vel pé de chanet [?] anglice dounhove vel wowell vel feldehove(C41) horsho, folefote, coltisfote oþer dinee [corr. dunehove](C42)
Lit. Alph., p.140, 24

Pes Urti

Rorippa nasturtium-aquaticum (L.) Hayek	Watercress

watercarse(C7)

Pes Vituli [p. vetuli]

Arum maculatum L.	Cuckoo-Pint
?*Portulaca oleracea* L./*sativa* L.	?Purslane

anglice ȝekestres(B2) wolvesfote(B4)
(pratinus minor) calvysfote(C1) pee de polyen, folefote [added .i. ȝekesters](C7) ȝekesterse(C10) anglice ȝokesters[sic](C16) cocowpyntell(C21)

Pes Vulturis

?

puttokesfot(C23)
Lit. Alph., p.140, 34

Petrophilon
Smyrnium olusatrum L. Alexanders
anglice stoncrop* vel stonmerch(B1)

Petrosillium [petrosillimum, petrosillum, petrosilinum, petrocillium]
Smyrnium olusatrum L. Alexanders, Horse-Parsley
Petroselinum crispum (Mill.) Nyman Parsley
persil(A3) gallice persyl(A4) persil(A11)
persil(B1) percill(B4) (petrosilion) oynouns*(B3) percel(B8)
persely (agreste minus) wyld persely(C1) anglice persell(C3)
percelly(C4) alexander-seed(C10) wilde persil(C14) (agreste) wild
percely (domesticum) .iii. species percely(C21) alisaunder-seed(C21)
alexandre-sede(C24)

Peucedanum
Peucedanum officinale L. Hog's Fennel
resteboef, kam(b)ock(B1) cammok(B3) cammoke(B4)
cammak(C2) dog-fenel(C5) anglice dogge-fenell(C6) cammok(C6)
houndfenel(C10) wormsede, swynesthistell*(C21)

Phales
?*Umbilicus rupestris* (Salisb.) Dandy ? Pennywort
penywort(C21)

Philantropos
Arctium lappa L. Burdock
borre(B3)
burre or clote(C4) clote(C6)
Lit. André I, p. 247 & II, p.196

Piganum [pigamen, pigamus, pigonnus, pigranum]
Ruta graveolens L. Rue
Thalictrum flavum L./*minus* L. Meadow Rue/Lesser Meadow Rue
anglice smale roude(A9) anglice smal rude(A12)
anglice rew(C3) wild rewe(C10) bellewort(C10) wild rue(C21)
bellewede(C24) bellewort(C27) wyld rwe(C27) pigell(C28) wild ruwe
vel smal rud(C37) wilde herbe(C39)
Lit. Alph., p.144, 24; N202

Pigmentaria
[*Melissa officinalis* L. Balm]
melilot*(C10)

Pigmentarium
Melissa officinalis L. Balm
anglice bawme(C6) baume(C36)

Pigris
Lactuca serriola L. 'Prickly Lettuce'
wild letuse(C21)
Lit. André p. 249f (picris) & II, p.199

Pigula [pigla, pigala, pidula]
Stellaria holostea L. 'Greater Stitchwort'
anglice stichewort(A2) pigel(A5 corrected)
gallice pigle, anglice stichewort(B2) pigle(B3)
pygul, bryddistong, stichewort(C1) briddestonge vel pigle(C2) (minor)
þe lasse pygull(C4)

Pilosella [pilocella, pilokella, pelucella]
Hieracium pilosella L. Mouse-Ear Hawkweed
gallice peluse[l](A5)
anglice lambestounge vel erswort(B2) peluette, mousere(B8)
gallice pilocelle, anglice lambestounge vel erswort(B2) mousere(B3)
peluet, pilosell(B4)
gallice pilosell, anglice hertworth* vel lambestong(C6) channet, mowshere(C10) chaunet, mousȝer(C11) gallice pilocelle, anglice lombestong vel erswort(C16, C18) hennewort*(C22) lampestunge .i. peilosele, heriwurt, mushere(C23) peluet, chenet, moushere(C24) chaunet, mouser(C27) gallice pulesete, anglice moushere(C39)

Pimpernella
Anagallis arvensis L. Scarlet Pimpernel
Pimpinella saxifraga L. 'Burnet Saxifrage'
Sanguisorba officinalis L. 'Great Burnet'
gallice pimpernole, anglice goldwort(B1) selfhele(B7) anglice houderugresse [sic](B9)
flewort(C2) gallice pympurnell, anglice selfhele(C6) pympernell, wlewrt(C7) pynpernele, selfhele, hervestplante(C8) selfhele(C24) saynt Mari golde*(C35)

Pinea [pine]
Pinus L. Pine
gallice pine, anglice pin-notte(A1)
anglice pyn-apples(C26) anglice pine(C29)

Pionia
Paeonia L. ssp. Peony
gallice pioyne(B2) pioyne(B3) gallice et anglice pione(B9)
(rotunda) þe more pyonye (femina) femel pyonye(C1) (rotunda, romana)
round pyony(C38)

Piper
Piper nigrum L. Black Pepper
Piper officinarum DC. White Pepper
(nigrum) gallice neyre peyvre(B2)
(nigrum) blac piper (longum) long pyper (album) whyt pyper(C1)

Pipera[?]
?*Stellaria holostea* L. ? 'Greater Stitchwort'
gallice lang de oysel, anglice stichewort(B1)

Pipla
?
yelwe bon, yelwe gold(C21)

Pira
Pyrus communis L. Pear
peres(B1)

Piretrum [peretrum, pelestrum, piretum, peletrum]
Anacyclus pyrethrum DC. Pellitory, Pellitory of Spain
Parietaria diffusa Mert. & Koch Pellitory-of-the-Wall
gallice peletre de Espayne(A1) peletre(A3) gallice peletre(A4, A9)
peletre savage(A5) gallice pelestre(A6, A8) gallice peretre(A12)
anglice [sic] pelestre de Spayne(B2) pelestre(B3) pelletere(B4)
mousere*(B5)
(domesticum, ortulanum, maius) peletre of Spayne (minus, agreste)
peletre(C1) anglice pelliter(C3) anglice peletre(C6) peletur of
Spayne(C14) (hyspanicum) þe gret pelleter(C31)

Pisa
Pisum sativum L. Pea

Cicer arietinum L. Chickpea
(alba, minor) pyse (nigra, maior) pese(C1)

Pissido
Pomegranate bark [*Punica granatum* L.]
le escette [corr. escorce] de pomme gernat(C22)
Lit. André I, p. 263 (psidia); Alph., p.147, 7 (persidia)

Pistacee
Pistacia vera L. Pistachio
festu(A3)
gallice fest[u] vel pistate(ch) [sic](C6) fruytes ylyche pynes(C10)
Lit. Alph., p.147, 9

Pisterimon [conf. of peristerion & periclimenon?]
?*Lonicera periclymenum* L. ? Honeysuckle
gallice cheverfoil(B7)
gallice cheverfoyle(C6)
Lit. cf. Alph., p.142, 4 & 7

Pitoniom [?]
?*Arum maculatum* L. ? Cuckoo-Pint
?*Dracunculus vulgaris* Schott ? Dragon Arum
dragans or nedderwort(C21)

Planta Leonis
?*Alchemilla vulgaris* L. ? Lady's Mantle
?*Fumaria officinalis* L. ? Fumitory
?*Taraxacum officinale* Weber, sensu lato? Dandelion
medilwort [corr. mesilwort](C21)

Plantago
Plantago major L. 'Great Plantain'
anglice plauntayne(B1) weibrede(B5)
planteyne or weybrede(C10)

Plantago Maior
Plantago major L. 'Great Plantain'
anglice weybred(A2)
plantayne, weybred(C1)

Plantago Minor
Plantago lanceolata L. Ribwort

anglice ribwort(A2)
rybwort(C1)

Platociminum [planto, platocimum, plantociminum]
Laserpitium siler L. Hartwort, Sermountain
Seseli montanum L. Sermountain
gallice silermontayne(A1)
gallice silermontaigne, sermountaigne(B2) comyn(B3)
suremontayne(C10) surmonteyne(C11) surmonteyn(C21)
Lit. André I, p. 255 & II, p. 202; N204

Podagra Lini
Cuscuta epilinum Weihe 'Flax Dodder'
doder(B3)
anglice dodyr(C3) anglice dodyr(C6) dogir[sic](C24)

Poetareon
a fern?
brake(C21)

Poleton
?*Oxalis acetosella* L. ? Wood-Sorrel
pisilotrum, anglice wodesure; et dicitur quandoque depilatorium vel panis cuculi(B1)
Lit. see **Psillotrum**

Poliarium
?*Mentha* L. ssp. ? Mint
wild horsmynt(C14)
Lit. cf. Alph., p.148, 39

Policaria [pollicaria]
Pulicaria dysenterica (L.) Bernh./ *vulgaris* Gaertn. Fleabane/'Small Fleabane'
?*Plantago indica* L. ? Fleawort
?*Inula viscosa* Ait.
anglice swynesgres*(A2) gallice policarii(A5)
pollicarie(C1) (minor) shabbewort(C10) (minor) scabbewort(C11)
romgresse(C14) policary(C21) scabwort(C21) (minor) skabworte(C24)
(minor) anglice scabyose(C26) (minor) scabwort(C27) foxglove*(C29)

Poligonia [polligonium, polligominum, pollegenia]
Polygonum aviculare L. — Knotgrass, 'Centinode', 'Swine's Grass', 'Sparrow-Tongue'
gallice centinodie, anglice swynesgras(A1) centinodie, swynesgras, gallice lange de mussun(A3)
centinodye(B1) gallice lange de musseroun vel centinodie(B2) sanguinaria, hencresses(B3) swynesgras(B3)
gallice poligenie(C5) gallice long de musseroun vel centenode(C16) gallice longe de musseroun vel centinodie, anglice swinesgrasse vel knottynggras(C18) anglice quichyn(C20) lange de mussun, cnotwort(C22)

Polipodium
Polypodium vulgare L. — 'Polypody'
Thelypteris dryopteris (L.) Slosson — Oak Fern
pollipodie(A3) gallice polipodie(A4)
polipodie .i. hocfern(B1) gallice pollipodie, anglice wodeke(B2) polypodie qe crest en le chene(B2) polipodie(B3) everfern(B5) vern apon ok(B8)
fern of okys(C1) everferne(C7) gallice polipode, anglice wodkey vel elferne(C16) gallice pollipodie, anglice wodekeye(C18) gallice pollipodie qe crest en le cheyne, anglice wodekeye(C18) anglice wodekei(C23) anglice wodekeye(C23) everfern growing in oke(C24) overfern(C28)

Politrichum
Adiantum capillus-veneris L. — Maidenhair-Fern
Asplenium trichomanes L. — 'Maidenhair Spleenwort'
gallice herbeneyt*(B1)
maidhere(C21) kex*(C29) romani vocant eam ayl savage*, anglici ramese*(C43)
Lit. André I, p. 257 (polytrichon) & II, p. 205

Polium [pollium]
?*Teucrium polium* L. — ? 'Mountain Germander', 'Poly' and other ssp. of Germander
.i. maner de polii(B2)
more pedelion*(C21) (agreste) ditany*(C21) anglice germandre(C26)
Lit. André I, p. 256 & II, p. 203; Alph., p.147, 31 & p.148, 12

Polliganos
= 'sanguinaria' = *Polygonum aviculare* L. here conf w. 'sanguinaria' =

Capsella bursa-pastoris (L.) Medic. Shepherd's Purse
sheperdispurs*(C21)
Lit. André I, p. 257 (polygonos) & II, p. 204 cf. **Poligonia**

Pollutritas [for politrichas]
Adiantum capillus-veneris L. Maidenhair-Fern
waterwort(C21)

Pomum Citrinum
Citrus limon L. Lemon
Citrus aurantium L. Orange
pumme de terre*(A3)
gallice poume de citre(B2)
pom citre(C11) gallice pumme de citre(C18) pomme de terre*(C22)
pome citrine(C24)

Pomum Silvestris
Malus sylvestris Mill. Crab Apple
crabbes(B3)

Pontarium
?*Teucrium scorodonia* L. ? Wood Germander, Wood Sage
ambrosie(B3)

Ponticum
Artemisia absinthium L. Wormwood
wermod(B3)

Populus
Polygonum aviculare L. Knotgrass, Swine's Grass
anglice swynesgras(A6)
swynesgras(B3)
þe bolas-tre*(C6)
Lit. Alph., p.148, 37

Porcelana [purcellana, porcellana]
Portulaca oleracea L. Purslane
gallice porce[l]ane(B2)
gallice porcelane(C5) gallice porcellane(C6) gallice purcellane, anglice assi letuse*(C20) purtzellan vel foelfote(C32)

Porrum [porrus]
Allium porrum L. Leek

treue gorlic(A10)
poret, anglice lekes(B1) anglice lekes(B2) gallice poreye(B2) lekys(B4) anglice leke(B7) porret, portyn(B8)
leek(C1) porrei, leke(C7) porret, leek(C21)

Portulaca
Portulaca oleracea L. Purslane
gallice portulane(A4) gallice porcelayne(A10)
pepoyl(B1) pé de poleyn(B5)
purslane(C1) portlake or purselane(C4) gallice porcelane(C5) folefote(C14) portulake, porsulane(C21) porcelline(C22) purswort(C30) helow(C42)

Postericum
?*Coriandrum sativum* L. ? Coriander
coriaundre(C21)

Potentilla
Valeriana officinalis L. Valerian
valeriane(B3)
herbe yve[?]*, lovewort, valerian(C21)
Lit. Alph., p. 8, 4

Prassium [prasion, prassinum, prassionum, prassio, prassum]
Marrubium vulgare L. White Horehound
Ballota nigra L.ssp.*foetida* Hayek Black Horehound
gallice marole, anglice horehunne vel horehunde(A1) maroil, horhunne vel horhunde(A3)
anglice horshune(B1) gallice maurul, anglice horhoune(B2) horhoune(B3) horshone(B4) anglice horhound(B7)
horehounde(C10) quidem anglice mogwort et quidem vocant anglice horhounde(C12) marou, horhound(C22) horoune, moryel(C27)

Pras(s)um
Allium porrum L. Leek
gallice porreus, anglice lek(A1) poret, leke(A3)
lek(C8)
Lit. André I, p. 260 (prason) & II, p. 207

Prassum
?*Marrubium vulgare* L. ? White Horehound
?*Ballota nigra* L. ssp.*foetida* Hayek ? Black Horehound

houndbery [for houndbene = houndbane?](C21)
Lit. Agnus Castus p. 235

Pratellus [pratella]
Melilotus officinalis (L.) Pall. 'Common Melilot'
Trifolium arvense L. Hare's-Foot
anglice honysokyll(C16) (agrestis) honysokyl(C21)

Priapiscus
Mentha pulegium L. Penny-Royal
Orchis L. Orchid
gallice polion(B1)
ravenleke(C21)
Lit. B II, 68 & Alph., p.146, 33; André I, p. 260 & II, p. 207; N205

Priapismus
Vitis L. ssp. Vine-Leaf
vynelef(B4)

Primula Veris
Primula veris L. Cowslip
Primula veris L. x *vulgaris* Huds. Common Oxlip
Primula elatior (L.) Hill Oxlip
gallice premerol(A2)
prymerole(C1) prymrose, oxlippe(C4) primerose, oxlyp(C21)

Primurella
Primula veris L. Cowslip
Primula veris L. x *vulgaris* Huds. Common Oxlip
anglice cuslop, sed primula veris gallice premerol et non sunt eadem herba sed multum assimilantur(A2)

Prosepinacia
Polygonum aviculare L. Knotgrass
centinodie, sparwetong, swynesgresse(C21)
Lit. André I, p. 261 (proserpinaca) & II, p. 208

Prosopus
Beta vulgaris L. Beet
anglice bete(B1)
Lit. cf. André I, p. 261 (prosopion) & II, p. 208

Prospinata
?*Cichorium intybus* L. ? Chicory, Wild Succory
cicori(B3)

Pruna
Prunus spinosa L. (fruit of) Sloe
Prunus domestica L. (fruit of) Plum Tree
gallice prunere, anglice ploumes(B2)

Pruna Damascena
Prunus domestica L. Plum
gallice prunes damascenis .i. neyr prunes(A4)

Prunella
Prunus spinosa L. Sloe
gallice pruneles, anglice sloes(A4)
gallice prunele dé bois(B2) anglice sloue(B7)
wodeslou(C1) heggetiken [for heggeþiken?](C2) gallice prunelles, anglice sloon(C5) gallice pruner dé boys, anglice wodsloue(C16)

Psillium
Plantago indica L. 'Psyllium Plantain', 'Fleawort', 'Fleaseed'
flewrt anglice(A5)
(domesticum) psylle (agreste, minus) wylde psille(C1) þe seede of an herbe(C10) seed of scabwort(C21) petrosillium, persly*(C21) herba pulicarum, mousere*(C32)
Lit. André I, p. 263 & II, p. 210

Psillotrum
Oxalis acetosella L. Wood Sorrel
depilatorium vel panis cuculi, alleluya, anglice stubwort(C12)
Lit. See **Poleton**

Pulegium
Mentha pulegium L. Penny-Royal
Thymus serpyllum L. Wild Thyme
gallice pulliole(A4) puliol(A11)
puliole, anglice brotherwort(B1) puliole(B1) anglice hulwort(B2)
anglice hillewort(C5) anglice holwort(C12) puliol, brotherwort(C15)
piliole(C27)

Pulegium Cervinum
Mentha arvensis L. — Hart's Penny-Royal, 'Corn-Mint'
herte puliol(C1) (p. cervinum/rurale) hertys pulial or felde pulial(C4) hert puleol(C21) feld pyliall(C38)
Lit. Alph., p.150, 22

Pulegium Martis
Origanum dictamnus L. — Dittany of Crete
Dictamnus albus L. — White Dittany
ditayne(C10) ditanndir(C11) ditandre(C21)
Lit. Alph., p. 51, 1 (note); André I, p. 264 & II, p. 211

Pulegium Montanum
Mentha pulegium L. — Penny-Royal
Thymus serpyllum L. — Wild Thyme
anglice pulliol muntayne .i. hillewort vel brotherwort(A1) pulliol muntayne .i. hilleworte vel brotherwort(A3)
(p.m. et gliconium) gallice puliole, anglice hulwort(B2) brothwort[sic](B4)
broþerwort(C1) hillewort, puliol mountayn(C4) anglice hylleworth(C6) puliole mountayne et bretherwort(C7) puleol montayn, hilwort, brotherwort(C21) lasse pyliall(C38)

Pulegium Regale
Mentha pulegium L. — Penny-Royal
Thymus serpyllum L. — Wild Thyme
gallice pulliol reale(A1) pulliol real(A3) anglice hulwort(A6)
alerworte, pulyall ryall(B4) alrewort(B8)
pulyol ryol(C1) alrewort, hulwort(C2) anglice puliall riall or chirchewort(C4) alyewort, pulyol real(C8) puleol real, cyrchewort(C21)

Pulegium Rurale
Mentha arvensis L. — 'Corn Mint'
feld puliol(C1) ffeld puleol(C21)

Pulicaria
Pulicaria dysenterica (L.) Bernh./ *vulgaris* Gaertn. — Fleabane/ 'Small Fleabane'
anglice flewort(B1)

Pulmonaria [pullimonaria, pullinovaria]
?*Galium cruciata* (L.) Scop. — ? Crosswort

?Pulmonaria officinalis L. ? 'Lungwort'
gallice croiser(A2) crosey(A8) galice croser(A9, A12)
anglice bellewort(B2) gallice cifer[?](B9)
belwort(C1) croyse(C2) croser(C3) croyser(C4) anglice bolwede(C5) anglice belwode(C6) croyser(C6) anglice croyse(C12) anglice bellewort(C18) carlok*, wild wortis*(C21) ffeldewort*(C22) bollewed(C23) madeneheyre(C34) bellewode vel scrophularia vel longwort(C36) gallice croyser(C37) caulis agrestis, anglice wilde caul*(C39)
Lit. Alph., p.151, 10

Quartifolium
?Paris quadrifolia L. ? Herb Paris
fetrewurd(B1)
ffeterwrt(C7)

Quercula Maior
Ajuga chamaepitys (L.) Schreb. Ground-Pine
Teucrium chamaedrys L. 'Wall Germander'
anglice rathele(A1) rathele(A3) q. maior et minor, quinquenervium, anglice ribwourt*(A6)
anglice ratele(B1) anglice ratele(B2) peniwort(B3)
ferþþ[i]nggras(C1) anglice ferthynggrasse, mederatle(C4) anglice ratyll(C6) mederatill(C10) meduratill, sterching-gresse, penygresse(C21) anglice germandre or hors clyre [corr. chyre](C41)
Lit. N206; Agnus Castus p. 242

Quercula Minor
Teucrium chamaedrys L. 'Wall Germander'
germandre(A3)
anglice ratele(B1) gallice germaundre, anglice medratele(B2) peniwort(B3)
germaunder(C1) anglice germaundre, hert[c]lovre(C4) anglice mederatyll(C6) germandr(C10) herteclowre(C32)

Quinquefolium
Potentilla reptans L. Cinquefoil
gallice quintefoyl, anglice five-levede gras(A1) quintefoyl, five-levyde gras(A3)
quintefoyle .i. fivelef(B1) gallice cincfoil(B2) synkefoyl, .v. lef(B8)
synkfoyl, fyf-levyd gras(C1)

Quinquenervia [quinquinervia, quinquenervium]
Plantago lanceolata L. Ribwort
gallice lanceley, anglice ribwort(A1) anglice ribwourt(A6) rubwort . . . launcelé(B1) launcilé(B4) anglice ribbe(B7) rybbewort(C4) kempis(C35)

Quinquilina
Plantago lanceolata L. Ribwort
lanceley, ribwort(A3)

Radiolum [radiola]
Polypodium vulgare L. Polypody
hocfern .i. polipodie(B1) everfern(B3)
everfern(C4) anglice everfarne(C12) everferne, rayele(C21) everfern(C32) radiol(C38)
Lit. B II, 40 & III, 80f; André I, p. 268 (radiolum) & II, p. 214

Radix
Raphanus sativus L. Radish
radysche(C1)
Lit. Alph., p.153, 16

Radix Herculis
Mercurialis perennis L./*annua* L. Dog's Mercury/ 'Annual Mercury'
anglice mercury or smerdock(C26)

Radix Minor
Genista L. ssp. Gorse
?*Calluna vulgaris* (L.) ? Heather
?*Sarothamnus* (L.) Wimmer ex Koch ? Broom
genest(C24)

Rami Cedri
Heartwood of *Juniperus communis* L., source of Juniper Tar Oil
gallice le fust de junevre(A1) le fust de junevre, junipir(A3)
gallice le frut de junipre, anglice grost oþe ye heth(B2)
gallice le frute de juniper, anglice gost on þe heth(C16, C18)
Lit. Alph., p.153, 22

Ramnus [rampnus]
Rhamnus catharticus L. Buckthorn
anglice hyndeberi(A1) griseley(A2) secundum aliquos hyndebery, secundum alios vero grysleys .i. spinokys(A3) anglice hyndeberi(A4)

þeveþorn anglice(A6) anglice þevetorn(A8)
anglice thevethorn(B1) anglice houndesberie* vel þevethorn(B2)
þeveþorn, grosiler(B3) þeveþorn(B5)
þewþron(C1) grysoler, thevethorn(C4) anglice hou[n]dysberyn* et secundum alios þevaþorn(C6) anglice thefberien(C12) grisiler, anglice þeveþorn(C19) hewethorn(C26) thefthorn(C28) griseler, þeveþorn(C31) wyt þorne(C35)
Lit. N207

Rapa
Brassica rapa L. — Turnip, Navew
navet, nap(A7)
rape(C1) rapis(C31)

Raphanus [raphanum]
Raphanus sativus L. — Radish
Armoracia rusticana Gaertn. Mey & Scherb. — Horse-Radish
Raphanus raphanistrum L. — Wild Radish, White Charlock
gallice raiz, anglice radiche(A1) gallice rayz, anglice radiche(A2) rayis, radiche(A3) gallice raice(A4, A5) gallice [sic] romeus, anglice radych(A6)
gallice raye, anglice radich(B1) gallice rayz, anglice radeke(B2) redich(B3) radyke(B4) radich(B5)
(albus) rafe (rubeus) red rafe(C1) radyke(C3) neppe ryall(C32) (albus) whytt rafe (rubeus) red rafe(C38)
Lit. N206

Rapistrum
Raphanus raphanistrum L. — Wild Radish, White Charlock
anglice wylde wyrte(A2) anglice cherloke(A2) cherlok, wilde wyrte(A3) ui[l]de uort vel carlos(A6) wilde wort vel karloke anglice(A9) anglice kammoc(A12) wilde wort vel anglice karlok(A12)
anglice kerlok(B2) carlok(B3) anglice wil wort vel carloke(B9)
(domesticum, maius) whyt piper (agreste, minus) cherlok(C1) anglice wylld worte vel carleke(C3) cherlok(C10) anglice cerlok(C12) (agreste) carlok(C21) sherlok(C21) (domesticum) whyt peper (agreste) cherlok, werlock, kerlock(C32)

Rapum Terre
Cyclamen hederifolium Ait. — Sowbread
erthnote or dilnote(C21)
Lit. André I, pp. 270f. & II, p. 216

Rasca Lini
Cuscuta epilinum Weihe 'Flax Dodder'
anglice doder(A2, A6, A9, A12)
anglice doddur(C4)

Rastena
Plantago lanceolata L. Ribwort
lanceolata seu quinquenervia, romani dicunt rasteine vel lancelé, anglici vero ribwort(C43)

Recopie [recorie, recoree]
Succisa pratensis Moench Devil's-Bit Scabious
recopie, forbytene(C10) recopee, forbetene(C11)

Regina [r. prati, r. pratis, regna]
Filipendula ulmaria (L.) Maxim. Meadowsweet, Meadsweet
gallice reyne, anglice medewort(A1) reyne, medewort(A3)
gallice reygne, anglice moderwort(B2) medwort(B5) reyne, medswot(B8)
medwort(C1, C2) medewort, medeswote, byrwort(C4) anglice moderworth(C6) anglice roynet, moderwort or medewort(C10) (r. herbarum) anglice medwort vel medsote vel lambwort(C12) gallice reigne, anglice moderwort(C16) gallice reyne, anglice moderwort(C18) (masculus) medewort, birwort, burnet* (famellus) ramse*, avence*(C21) moderwort(C21) medwort(C35)

Resta Bovis [rasta b.]
Ononis repens L. Restharrow
restebeof, cammok(A3)
gallice reste de beste [corr. bef], anglice cammoc(B2)
cammok(C1,C4) gallice rest de befys, anglice cammok, hareberde(C6) anglice camoke(C18) cammok, whyn(C21) reste de boef, cammok(C22) kamok, reste de bove(C32) cammok vel wynnes(C36) gallice restyuef[sic], anglice restelbowe(C39)

Resuris
Rheum rhaponticum L. Rhubarb
rubarbe(C24)

Reubarbarum [raubarbarum]
Rheum rhaponticum L. Rhubarb
gallice reubarbe(A1) reubarbe(A3) gallice reubarbe(A4)
gallice et anglice rubarbe(B2)
radix est herbe reubarbe(C1) anglice reubarbe(C26)

Reuponticum [reuponticus]
[*Rheum rhaponticum* L. Rhubarb]
?*Sarothamnus scoparius* (L.) Wimmer ex Koch ? Broom
(agreste) anglice brome(C3) reed rose*(C21)
Lit. Alph., p.157, 6

Rixio
?*Geranium molle* L./ *columbinum* L. ? 'Dove's-Foot Cranesbill'/ 'Long-Stalked Cranesbill'
sperge, mawort(C21)
Lit. Agnus Castus pp. 247f.

Rix Naceus
?*Ulex europaeus* L. ? Furze, Gorse
gersk, ffyrces(C21)

Robelia
Pisum sativum L. Pea
Lathyrus L. ssp. Vetchling
þe comune pese(C10) þe comon pise(C11) the comon pes(C21)
Lit. Alph., p.155, 4

Rodina
?*Osmunda regalis* L. ? Royal Fern
?*Polypodium vulgare* L. ? 'Polypody'
overfern(C28)

Rodon
Rosa L. spp. Rose
anglice red rose or dobelle rose(C26)

Rogotea [for gerontea?]
?*Senecio vulgaris* L. ? Groundsel
groundswille(C21)

Romeus
Raphanus sativus L. Radish
anglice radiche(C15) raf(C21) anglice radick or radich(C26)
Lit. Alph., p.155, 2–3

Roncea
Rubus fruticosus L. Bramble

brembel(C2)
Lit. André I, p. 276 (runcia buttunaria)

Rorastrum
Rosa canina L. Dog Rose
Bryonia dioica Jacq. Red and White Bryony
anglice dog-rose(C29) bryoine, bryan or wyld nep(C31)

Ros Ciriacus
Althaea officinalis L. Marsh Mallow
ye flour of malwes(C21)
Lit. Alph., p.155, 19; N207f.

Ros Marinus
Rosmarinus officinalis L. Rosemary
rosmaryne(A5)
rosmaryne(B4)
gallice rosemaryn, felde-mader(C6) rosemary, fel[d]-madre(C21)
rosmaryne, feld-mader(C32)
Lit. B III, 88 (feldm[to]dere)

Rosa ['simpliciter est rosa rubea']
Rosa canina L. Dog Rose
rose(A3)
rose(B1)
heopebrumbel(C19)

Rosa Canina
Rosa canina L. Dog Rose
anglice hu[p]ebremyll(C4) huppbrer, eglentyne(C7) the blosson of whitthorn or whithorn(C26) houndesrose(C33, C37)

Rosa Marina
?
romani vocant eam rosemarin, anglici winter-rose(C43)

Rosa Passa [rosophassa]
?*Sonchus oleraceus* L. ? Milk- or Sow-Thistle
anglice hareberd(B7)
milkethistill(C4) hareberd(C8) harebert(C36)
Lit. Alph., p. 21, 11; 154, 39

Rostrum Porcinum [rostum p., r. porci, rastrum porcinum]
Sonchus oleraceus L. Milk- or Sow-Thistle
anglice sugþistel(A2) soweþistel(A3)
anglice southystel(B2) swynthistell(B4) souueþistel(B5) gallice scariole, anglice swynethistel(B9)
sowþystel(C1) anglice swynethystyll(C3) anglice milkethistyll(C4) sou(re)thistyl(C22) thowistill(C24) milk-þistil or swynisþistil(C31) camhoc(C35)

Roticis
Aristolochia clematitis L./ *rotunda* L. Birthwort
myke(C21)

Rubarbum [rubbarus]
Rheum rhaponticum L. Rhubarb
?*Sarothamnus scoparius* (L.) Wimmer ex Koch ? Broom
rubarbe(B1)
anglice rubarbe(C6) þe rote of þe leste geneste(C11) ye rote of lasse genest(C21)
Lit. See **Reuponticum**

Rubea Maior [rubeum]
Rubia tinctorum L. Madder
gallice warence, anglice madir(A1) gallice warence(A2) warence, mader(A3) gallice warence, anglice madir(A4) gallice warence, anglice mader(A6) mader(A7) gallice warance, anglice mader(A8) gallice warence(A9) anglice madyr(A10) gallice waurence(A12)
maddir(B1) gallice warence, anglice mader(B2) mader(B3)
madyr, warance(C1) madur, warance(C4) (r. agrestis) wild madre(C21) chete* [for clete], heirif*(C22) wild mader(C32) þe more madyr(C38)

Rubea Minor
Galium aparine L. Cleavers
anglice heyreve(A1) anglice heyryve(A2) clite, heyrif(A3) anglice heyreve(A4) gallice [sic] ayrive(A6) anglice hayrive(A8) anglice harive(A9) anglice heyrive(A12)
herbe tenaunte, [c]leveregras(B1) gallice [sic] clivere, anglice hayryve(B2) heyryve .i. clyvere(B3) anglice hayrif(C26)
anglice cliþe(C1) anglice hayrehefe(C3) clyver, hayrouve(C4) hayrif, anglice clyþrone(C10) anglice heggecliveren(C12) anglice hereeve [sic] vel cokwort*(C34) anglice heyreve vel cokewort*(C37) clyte(C38) hayreff, clyver oþer aron*(C42)

Rubea Parella
Rumex sanguineus L. — 'Red-Veined Dock'
rouge parelle et red dok(C7)

Rubus
Rubus fruticosus L. — Bramble, Blackberry
anglice bremel vel bremyl(A1) bremyl(A3) anglice bremyl, gallice runsun(A4) bremel(A5)
bremblеs(B3)
brembelbossshe(C2) (cime rubi) brembelcroppes(C2) gallice ronce, anglice bremyll(C4) brembelys, wote ye wyll þat þer be many brembyll, but whan rubus simpilli is put, as þe brembyll þat beryth blakeberys it shal be undirstoud(C11) anglice bremmel bosche(C12) blakbery, brembill(C24) (albus) bremb(r)ell (rubeus) anglice þe rede breris(C28)

Rudam
Ruta graveolens L. — Rue
rude(B3)

Rumex
Rumex L. spp. — Dock
moche dok(C21)

Runceus
Cuscuta epilinum Weihe — 'Flax Dodder'
doder(B3) anglice doder(B9)

Ruscus
Ruscus aculeatus L. — Butcher's Broom
Ulex europaeus L. — Furze
knehullen(B1)
firsyne(C6)
Lit. B III, 55f; André I, p. 277 & II, p. 221

Ruta [r. bassarum]
Ruta graveolens L. — Rue
Thalictrum flavum L./*minus* L. — 'Meadow Rue'/ 'Lesser Meadow Rue'
gallice rue, anglice smale-rede [corr. -rude)(A1) rue(A3) anglice rue(A4)
rue(B1) gallice et anglice ruwe(B2) (agrestis) anglice wilde rue(B2)
rue(B3) rewe(B4)
ruwe (agrestis, montana) wyld ruwe(C1)

Ruta Agrestis
[*Thalictrum flavum* L./*minus* L. 'Meadow Rue'/ 'Lesser Meadow Rue']

[*Peganum harmala* L. 'Wild Rue']
[*Ruta montana* L. Rue]
anglice woderofe*(C4, C31)

Rutialis
Agrimonia eupatoria L. Agrimony
egremony, goseclite(C21)

Sacrefolium
?*Sempervivum tectorum* L. ? Houseleek
[sin]full in englishe tonge, yn romayne jubarbe(C42)
Lit. B III, 207f.

Sagapinum [segapinum; headword erroneously conflated with sansucus entry]
?*Majorana hortensis* Moench ? Sweet Marjoram
mochworte*(C11) mugwort*(C21)
Lit. cf. Alph., p.160, 8. See **Sigapium**

Salgia [salvia, salgium]
Salvia officinalis L. Sage
Teucrium scorodonia L. Wood Germander, Wood Sage
gallice et anglice sauge(B1) gallice et anglice sauge(B2) sauge(B4, B5, B8)
sage (agrestis maior) wyld sage(C1) (agrestis) anglice hyndhale or wild sage(C4) (agrestis) hyndehale, ambrose(C4) (agrestis) anglice wild sauge vel hyndleddir [sic](C16)

Saliunca
Ulex europaeus L. Golden Gorse, Furze, Broom
Eryngium campestre L. Eryngo
Valeriana celtica L. Celtic Spikenard
gallice jaune, anglice firsis(A1) jaune, fyrsis(A3) anglice calketrappe(A6) anglice calketrappe(A8, A12) anglice kalketrappe(A9) calketrappe(B2) anglice calketrappe(B7) gallice calketrape(B9) kalketrappe(C1) anglice spik vel calketrap(C5) foxglowe*(C6) anglice spykenell, kalketrappe(C6) gallice spik seltik, anglice calketrap(C16) gallice spik celtik, anglice calketrappe(C18) kalletrape, breres in valeyes that maketh folke fall(C33) kalketrappe, breres in a valey þat maken folk

fallen(C37)
Lit. Alph., p.159, 4 (notes); B III, 189 (popig); André I, p. 279 & II, p. 224. See **Salmenta**

Salix
Salix L. ssp. — Willow
Salix viminalis L. — Osier, Withy
gallice seuce, anglice wyly [corr. wythy?](B2)
anglice whythy(C6) a wythy-tree(C10) welow-tre(C11) gallice seuce, anglice wythy(C18) a salwe-tre(C21) welew-tre(C24) (rubea et alba) þe wiþy þe rede and the whyte(C31)

Salix Marina
Vitex agnus-castus L. — Agnus Castus
sewyþi(C5) anglice sorwyþy(C6)
Lit. André I, p. 279 & II, p. 224

Salmenta [salmita, salamita; all errors f. saliunca]
Ulex europaeus L. — Furze
Eryngium campestre L. — Eryngo
Valeriana celtica L. — Celtic Spikenard
Ononis repens L. — Restharrow
calketrappe(B3) wanteloy*, voxglove*(B8)
karterappe [corr. karltrappe?](C3) kalketrap(C7) anglice kalketrappe(C15) spik, calktrappe(C21) restebowe(C29)
Lit. cf. Alph., p.160, 40 (salmentica)

Salpo
Dipsacus fullonum L. — Teasel
Sonchus oleraceus L. — Milk- or Sow-Thistle
southistil(C21)
Lit. André I, p. 279 (salpho)

Salvia Romana
Teucrium scorodonia L. — Wood Germander, Wood Sage
ambrosie(B3)

Salvia Silvestris
Teucrium scorodonia L. — Wood Germander, Wood Sage
hindehale(A2) sage savage, hyndehale(A3)

Sambucica [see Saliunca]
calketrap(C4)

Sambucus [sambuca]
Sambucus nigra L. Elder, 'Bour Tree'
gallice suy, anglice ellir-tre vel bur-tre(A1) su, hellerne(A3) anglice hellerene(A6) anglice heller(A8) elren(A11) gallice sambu, anglice bure-tre(B9)
ellen-tre(C2) anglice bur-tre(C3) anglice heldrene(C15) gallice scard, anglice ellerne-tre(C18) anglice elleberye(C22) eleer(C24) eldren(C28) ellerne or eldyr(C31) hilder-tre(C35)

Samsucus [samsucon, sampsucon, samsuqus, sansucus]
Majorana hortensis Moench Sweet Marjoram
conf. with sambucus (ebulus)
hundefenyl(A3)
sanksu .i. kyngeswort*(B1) maithe(B3)
majorane, muchewort(C10) walwort*(C21)
Lit. André I, p. 280 (sampsuchum) & II, p. 225; B III, 68 (cyningeswyrt) & 80 (ellenwyrt)

Sanamunda [sanamenda]
Geum urbanum L. Wood Avens
harfot, avence(A3)
avence(B1) avence(B3)
anglice avance(C3) avence, harefote(C21)

Sandalis [sandalus]
Santalum album L. White Sandalwood
Pterocarpus santalinus L. Red Sandalwood
anglice saundres(B2) saundyrs red and whyt(C31)
Lit. Alph., p.161, 31

Sandalum
?*Caesalpinia* L. ssp. ? Brazilwood
brasillum, anglice bracell(C29)

Sandix [scandix, sandax]
Isatis tinctoria L. Woad
Rubia tinctorum L. Madder
gallice wayde(A1) bleu woode(A3)
gallice warence, anglice madir(B1) wod, gallice wayde(B1) gallice wayde,anglice wod(B2) mader(B5)
(maior) wode (minor) wyld wode(C1) anglice wadde(C3) blew wood(C22) (silvestris) wild wode(C32) wold(C36)

Sandonicum
Artemisia maritima L. 'Sea Wormwood'
?*Santolina chamaecyparissus* L. Lavender Cotton
sewarmode(C10) seewormode(C11) anglice warmode that beryth wormsede(C26)
Lit. André I, p. 281 (santonica herba) & II, p. 226

Sanguinaria [sagliaria]
Capsella bursa-pastoris (L.) Medic. Shepherd's Purse, Sanguinary
Polygonum aviculare L. Knotgrass
gallice sanguinarie vel oumener(A1) sanguinarie, stanche(A3) gallice sanguinarie, anglice oumener(A4) stanche(A11)
staunchewort(B1) gallice burse de pastour, anglice houuemenere(B2) hennecresses(B3) sa[n]guinarie, bleddyrwort [error for bleddewort?](B8)
(maior) millefolium magis, bursa pastoris (minor) blodwort, sanguinarie(C1) (minor) polligonos, gallitricum, sanguynary(C4) anglice staunchebIode vel holkemeny(C5) anglice stanch vel sanguinare(C18) bloodwort, sheperdispurs, croweleke*, harebell*(C21) (maior) sheperdispurs, tothwort(C21) staumche [sic](C22) hencresse(C29) symonwort, bloodwort(C32)
Lit. André I, p. 281 & II, p. 226

Sanguis Draconis
Red gum or resin, 'cinnabaris', from a variety of species, esp.
Daemomorops draco Blume Dragon's Blood
Dracaena draco L. Dragon Tree
gallice saunk de dragune(A1) sank de dragoun(A3)
sank-dragon(C10, C24)
Lit. N209

Sanicula [saniculus, saniculum, saniclum]
Sanicula europaea L. Sanicle
gallice sanicle, anglice siwariswort vel wodemerche(A2) senicle, wodemerche(A3) gallice senikyl(A5)
gallice sanicle, anglice wodehache(B2) sanicle(B3) senekyll(B4) senigle(B5) canicle, wodmerche(B8)
sanpcle, wodemerche(C1) senigyll(C3) sanycul, wodemerche(C4) gallice sanicle, anglice wodeworte(C6) sanicle, wodmerch vel smerdesworthe(C16) sanicle, wudemirche(C23) wodewort(C36)

Santo [?]
Sinapis L. Mustard (seed)
mustardseed(B4)

Sapa
?Saponaria officinalis ? Soapwort
savoyne(A7)
Lit. Alph., p.159, 5 (sapo)

Saponaria [soponaria]
Saponaria officinalis L. Soapwort
?Saponaria ocymoides L. ? Rock Soapwort
sovonere, crousope(A3)
crowsope, savenere(B4)
(maior) crowsope, saponere (minor) lasse saponere(C1) herba fullonum, anglice tasyll(C3) anglice crowsope or brownewort(C4) savener, crowsope, netebroun(C10) crowsope, tasill(C21) savuner, crousope(C22) cavaner, crowsope(C24) crowsope or fane*(C29) (minor) lass krowsope(C41)
Lit. Britten & Holland Dict., p. 68 record brown net and brownwort in ref. to spp. of Scrophularia

Saracenica [for mentha saracenica?]
[*?Chrysanthemum balsamita* L. ? Costmary]
woderove*(B3)

Sarallon
Lactuca serriola L. 'Prickly Lettuce'
Lactuca virosa L. Lettuce
endive, hors-thistil(C21)
Lit. André I, p. 282 (sarralia) & II, p. 227 (sarracia)

Sarcocolla [sacricolla]
Astragalus fasciculifolius Boiss. Gum Sarcocolla
Bryonia dioica L. White Bryony, Wild Nep, Wild Vine
Agrimonia eupatoria L. Agrimony
gallice brioine, anglice wylde nepe(A1)
gallice et anglice sarcocolle(B2) egrimoyne(B3)
gallice sarcocole(C5) gallice sartacok(C6) the gum of a tre(C21)
Lit. André I, p. 281 & II, p. 227

Sarmentum Fraxini
Fraxinus excelsior L. Ash-Shoot
scryding of ash(C7)

Satirion [satureon, saturian, saturion, seterion]
Orchis L. ssp. esp. *Anacamptis pyramidalis* (L.) Rich. Orchis/ 'Pyramidal Orchid' and *Orchis mascula* (L.) L. Early Purple Orchid
Arum maculatum L. Cuckoo-Pint
?*Endymion non-scriptus* (L.) Garcke ? Bluebell, Wild Hyacinth
anglice stondilgourde(B2) standilgrus(B4) yekesters(B5)
ʒekus, revenyslekus (minor) crowlekus, harebell(C1) ravensleke(C4) anglice standelwort(C5) anglice standworth(C6) standulgrasse(C7) standelgusse, ʒekesters(C8) standelgres(C12) anglice s[t]andegourde(C16) ravenleke(C21) cokkocpintel(C24) cuckowpintill(C28) gandergas(C31) rammeslek(C32) anglice hoggspyntyl(C39) (maior) anglice ʒeke[st]ers, [s]chandel-welk(C41) (maior) scandilgose oþer ʒekesteris oþer skandylwekys(C42)
Lit. B II, 68; André I, p. 282 & II, pp. 227f; Alph., p.158, 19 & 36; Agnus Castus p. 202, 7; N210

Satucella
?*Symphytum officinale* L. ? Comfrey
anglice consounde(C4)

Satureia [saturegia]
Satureia hortensis L. Summer Savory
Satureia montana L. Winter Savory
gallice sarré, anglice sathre(A1) galice sateré(A2) sauterel(A2) sarré, saturé(A3) saturele, anglice saturay(A5) gallice savoroye(A6) gallice saveray(A9) gallice saveroy(A12)
sathereye(B1) gallice et anglice satureye(B2) savoreie(B3)
saverey (agrestis minor) wyld saverey(C1) saverey(C4) gallice sair vel saturier(C5) gallice sarre vel saturae vel sache(C6) savereyn(C24) savoré(C35)
Lit. André I, p. 282 & II, p. 227

Savina
Juniperus sabina L. Savin
savine(A3) gallice savine(A4)
savine vel bractes(B1) gallice et anglice savine(B2)
savyne(C1) anglice sawyn(C3) romani vocant eam savyne, anglici buterbesome(C43)

Saxifraga [saxifragia, saxifragium]
Saxifraga L. ssp. Saxifrage
Pimpinella saxifraga L. Burnet Saxifrage

Asplenium adiantum-nigrum L./ *trichomanes* L.	'Black Spleenwort'/ 'Maidenhair Spleenwort'
Ceterach officinarum DC.	Rusty-Back Fern

gallice saxifrage(A1, A4) saxifrage, stonbrekke(A3) stonwrt(A11)
saxifrage(B1, B4) gallice saxifragie, anglice stenbrok vel papiwort [= pimpinella](B2) papwort(B5)
saxfrage (minor) þe lasse saxfrage(C4) gallice saxifragie, anglice papeworth(C6) saxifrage, brekestone(C21) saxifrage, pappewort(C22) pappewort, stonwrt(C23) saxfrage, brekestone(C31)
Lit. André I, p. 283 & II, p. 228

Scabescella [scabascella]

?*Anagallis arvensis* L.	? Scarlet Pimpernel
?*Pimpinella saxifraga* L.	? 'Burnet Saxifrage'
?*Sanguisorba officinalis* L.	? 'Great Burnet'

saywort, hervystplaunte(C8) saywort, hervestplant(C13)
Lit. see **Pimpernella**

Scabiosa

Knautia arvensis (L.) Coult.	'Field Scabious'
Scabiosa columbaria L.	'Small Scabious'
Succisa pratensis Moench	Devil's-Bit Scabious
Centaurea iacea L./ *scabiosa* L.	'Lesser Knapweed'/ 'Greater Knapweed'
Inula helenium L.	Elecampane, 'Scabwort'

gallice scabiose, anglice shabbewort(A1) scabwort(A3) anglice shabwort(A4)
gallice scabiouse, anglice scabewort(B2) deloroune, scabius(B4) scabiouse(B5)
(maior) scabiose (minor) lasse scabiose(C1) gallice dolurune, anglice scabiose . . . anglice scabyous(C3) anglice scabwort (minor) .i. matefelon(C6) deloroune, scabyose (minor) matefelon(C8) anglice scabiouse, gallice dolurune*(C12) matfeloun(C15) scabous(C21) scabwort(C22) horsfartis(C35)

Scalones [escalones]

Allium ascalonicum L.	Shallot
Allium cepa L.	Onion

scalouns(A3)
scalonis(C1) scalons(C21)

Scamonia

Convulvulus scammonia L.	Levant Scammony

gallice et anglice scamoyne(B7)
gallice et anglice scamoine(C5) gallice et anglice scamaunce vel scaverole(C6)
Lit. André I, p. 283 & II, p. 228

Scandula
Lolium temulentum L. Darnel
darnel(C6)
Lit. André I, p. 283 (scandala, scandulaca, scandulacium) & II, p. 229

Scariola [sclariola, slariola]
Lactuca serriola L. 'Prickly Lettuce'
Sonchus oleraceus L. Sow- or Milk-Thistle
gallice scariolei vel endivie vel lettuse savage(A1) scarioley vel endive vel letus savage vel felde-letuse(A3)
sclarie(B1) endyve(B3) slarie(B3)
sariole, harde-hakke(C1) anglice shariol .i. mylkethistill(C4) (silvestris) wilde letuse(C10) anglice saveraye*(C20) scarioll, hardhak(C21) wilde letuse, slepewort(C22) schariole(C32)

Scatumcellus [satuncellus, scatocella, statuncellus, scatuella, scatumcella]
Umbilicus rupestris (Salisb.) Dandy Pennywort, Navelwort
gallice denarole(A1) deinerol, peniwort(A3) anglice peniwort, gallice denerole(A4)
gallice denerole, anglice peniwort(B2) peniwort(B3) gallice denariole, anglice peniwort(B7)
gallice denaroil, anglice penyworth(C6) growndeswyllys* vel potius penyworte(C6) gallice denerole, anglice penywort(C18) deinerol, penywort(C22)
Lit. Alph., p.163, 25

Scelerata (Herba)
?*Ranunculus bulbosus* L. ? 'Bulbous Buttercup'
Ranunculus sceleratus L. 'Celery-Leaved Crowfoot'
?*Anthriscus sylvestris* L. ? Wild Chervil
anglice cloftong(C16) anglice cloftunge(C18) cerfoile, crowfote(C21)
Lit. André I, p. 284 & II, p. 229

Sceleratum
Lolium temulentum L. Darnel
popul(C14)

Scicida
Bryonia dioica Jacq. Bryony
?*Ecballium elaterium* A. Rich. ? Squirting Cucumber
brionie, wylde nepe(A3)
gallice brione, anglice wylde nep(B2)
wilde nep(t)e(C22)
Lit. Alph., p.170, 1; N167; see **Siccida**

Scilium
?
anglice clyll(C3)

Scilla
Iris L. ssp. Gladden
Gladiolus L. ssp. Gladiolus
glaole, gladene(B1) gladen(B1)
Lit. André I, pp. 284f. & II, pp. 229f; B III, 113

Sclareia
Salvia sclarea L. Clary
Salvia pratensis L. 'Meadow Clary'
Salvia horminoides Pourr. Wild Clary
sclarie(B1)
Lit. André I, p. 285 & II, p. 230

Scolopendria [scolopendrium]
Phyllitis scolopendrium (L.) Newm. Hart's-Tongue Fern
gallice cerlonge, anglice hertistong(A1) cerflang, hertestounge(A3)
gallice cerflong(A4)
serflang(B1) gallice cerlange, anglice hertestonge(B2) cerlonge(B3)
herttestonge(C10) pié de coc(C17) hert-tonge(C28)
Lit. André I, p. 285 (scolopendrion) & II, p. 230

Scolopendula [conf. w. calendula?]
?*Calendula officinalis* L. ? Pot Marigold
rode(B5)

Scopa
Betula L. ssp. Birch
birche(C21)
Lit. André II, p. 231

Scopa Regia
Hypericum perforatum L. St. John's Wort
anglice seint Johanhis wort(A1) Johanhiswort(A3) herbe Johan(A3)
herbe Jon(B3)
seyn Jonys wort(C10) herb John(C21)
Lit. cf. André II, p. 231

Scordion [scodion, scoridion]
Allium ursinum L. Ramsons
Allium vineale L. Crow Garlic
Endymion non-scriptus (L.) Garcke Bluebell, Wild Hyacinth
gallice aile savage, anglice wilde gorlek(A1) ail savage, wylde gorlec(A3)
gallice aylle savage, anglice garlek(B2) croulek(B3) crowegarlek(B5)
crowgarleke(C4) harebolle(C10) harebell(C11) ramse, crowleke(C21)
harebell(C21) harbell(C24)
Lit. André I, p. 286 & II, p. 231

Scrophularia [scrofilaria]
Scrophularia nodosa L. Figwort
?*Ranunculus ficaria* L. ? Lesser Celandine, Pilewort
scrofilarie(B3)
anglice boonwort(C19)

Segetalis
Gladiolus L. Gladden
gladyn(C21)
Lit. André I, p. 288 & II, p. 233

Selinon
Apium graveolens L. Wild Celery, Smallage
smalache(C21)
Lit. André I, p. 288 & II, p. 234

Selinum
Apium graveolens L. Seed of Wild Celery
anglice merchseede(C10) mercheseed(C21)
Lit. André I, p. 288 (selinon) & II, p. 234

Semen Rapistri
?*Lepidium sativum* L. ? Garden-Cress Seed
cres-sed(C11) touncras-sede(C24)

Semen Veneris
Sonchus oleraceus L. Sowthistle
soweþistil(A3)

Semperviva
Sempervivum tectorum L. Houseleek, Sengreen
jubarbe, singrene(A7)
gallice jubarbe, anglice sinegrene(B2) syngrene(B3) singrene(B5)
anglice sengrene, jubarbe(C1) anglice synegrene(C10) selgrene, jubarbe, housleke(C21) anglice synchoun [corr. syngrene?](C29)
Lit. André I, p. 289 & II, p. 235

Senacium [senacio, senicium, senecium, cenacon]
Rorippa nasturtium-aquaticum (L.) Hayek Watercress
?*Lepidium sativum* L./ *campestre* (L.) R.Br. ? Garden Cress/Pepperwort
gallice faverole, anglice waterkersen(B2) anglice wilde cressus(B7)
anglice wyld cresse(C6) watercars(C10) watercresse(C21) watercarsis or welcarys(C31)
Lit. Alph., p.165, 24 (senacio)

Senecio [senecion, senacion, senecium, seniscon, seniscum, senicon, senasion, sanasion, senexio]
Senecio vulgaris L. Groundsel
Cnicus benedictus L. 'Holy Thistle'
gallice sceneccun, anglice grundyswylye(A1) groundyswyly(A3) anglice grundeswelye(A6) anglice quichin*(A8) anglice synechon(A12)
gallice seneson, anglice grundeswell(B1) gallice cardoun beneit, anglice grundswele(B2) groundeswelie(B3) groundeswylie(B5)
groundswelie(B8) senachon, grondswal(B8)
grundyswely(C1) growndeswilie(C2) anglice groundeswilie, wallcrasse(C12) groundswille, sencion(C21) grounsaile(C24)
grondeswall(C29) gallice synichon, anglice grownswale(C39)
groundewyly or swython [sic](C41)
Lit. André I, p. 289 & II, p. 235

Senellio
?
chekinmete(C28)

Senicon
?*Sempervivum tectorum* L. ? Houseleek

cardus benedictus, anglice synthym [corr. syngren?] vel howseleke(C3)
Lit. André II, p. 235 (senecio)

Sentonica
Artemisia maritima L. 'Sea Wormwood'
sewermot(A3)
Lit. see **Sandonicum**

Sepha
Hyssopus officinalis L. Hyssop
ysope(C21)

Septemnervia
Plantago major L. 'Great Plantain'
weybrede, planteyn(C21)
Lit. André I, p. 290 (septinervia) & II, p. 235

Seralia
Lactuca serriola L. 'Prickly Lettuce'
endyve(B3)
Lit. André I, p. 282 (sarralia) & II, p. 227 (sarracia)

Serone
?*Betonica officinalis* L. ? Betony
betayn, bisshoppiswort(C21)

Serpentaria
Dracunculus vulgaris Schott Dragon Arum
Polygonum bistorta L. Bistort, Snakeweed
gallice serpentarie vel dragance(A1) serpentarie vel dragance vel serpentine(A3) anglice serpentarie(A4)
serpentarie, anglice nedderwort(B2) dragaunce(B3, B5)
gallice serpentaria [sic], anglice neddergyrse(C18) serpenture [sic](C22)
Lit. N212

Serpentina [serpentinarea]
Dracunculus vulgaris Schott Dragon Arum
Polygonum bistorta L. Bistort, Snakeweed
nadderwort(B1)
neddregresse(C14) neddre[-wort], dragance(C21)

Serpillum [serpillus, scupillum]
Anacyclus pyrethrum DC. Pellitory of Spain

Allium porrum L. Leek
gallice peletre(A1) peletre(A3) pelestre(A7)
perestre, anglice petir vel brodelek(B1) gallice, anglice pelestre(B2) pelestre(B3)
peletree(C10) anglice peletre(C22) wylde peleter or þe lityl pelleter(C31) anglice peletre vel peterwort(C39)
Lit. B III, 36 (bradeleac)

Serucium [for strucium?]
?*Brassica* L. ssp. ? Cabbage
wylde cawel(C22)

Setacul
Eryngium maritimum L. Sea Holly
gallice yringis, anglice se-holine(A1) yringys, se-holin(A3) gallice yringis, anglice se-holine(A4)
yringi, anglice se-holy(B2)
se-holine(C4) yrynges, se-holine(C10)
Lit. N211

Sibillum
Anchusa arvensis (L.) Bieb. Bugloss
lang de bef, oxtung(C21)
Lit. André I, p. 292 (sibilla) & II, p. 238 (sibissa)

Siccida
Bryonia dioica Jacq. Bryony
?*Ecballium elaterium* A. Rich. ? Squirting Cucumber
wilde nepe(B7)
anglice wilde nepe(C4)
Lit. Alph., pp.170 & 173; N167; see **Scicida**

Sicla [bete sicle]
Beta vulgaris L. [var. Cicla] Beet
betes(C10) gallice, anglice betis(C16) bete(C17) betis(C21)
Lit. André I, p. 53 (beta Sicula) & II, p. 35

Sidia [psidia]
skin of pomegranate [*Punica granatum* L.]
gallice la escorce de pome gernete(A1) gallice le escorce de pome gernet(A4)
Lit. André I, p. 293 (sidium); N205f.

Sigapium [headword erroneously conflated with next entry, Samsucus]
?*Majorana hortensis* Moench — Sweet Marjoram
?*Origanum vulgare* L. — Marjoram
mechewort(C24)
See **Sagapinum**

Sigillum Salomonis
?*Paris quadrifolia* L. — ? Herb Paris
Polygonatum multiflorum (L.) All. — Solomon's Seal
tuttesayn(C4) tuteseyn et tutsayne(C7) Salamonys seale(C21)
Lit. Alph., p.169, 7

Sigillum Sancte Marie
Polygonatum multiflorum (L.) All. — Solomon's Seal, 'St. Mary's Seal'
seynt Marie flour(A3)
anglice seynte Marie flour(B2)
seynt Mary sele(C4) anglice seynt Mary flour(C6) seynt Mary seale(C21)
seynt Marie sel(C32) rensis[?](C35) seynte Mary sede [corr. sele](C41)
seynt Mary seall oþer godhis-hond(C42)
Lit. Alph., p.169, 3

Siler Montanum
Siler Montanum Crantz — Sermountain
Laserpitium siler L. — Sermountain
silermontayne(A3) gallice silermontayne(A4)
sirmontayne(B3)
seyrmonteyn(C1) anglice sirimontane(C3) anglice surmontayne(C6)
Lit. André I, p. 293 & II, p. 239; N213

Sileum
Apium graveolens L. — Seed of Wild Celery
merchseed(C10) merchsede(C24)
Lit. André I, p. 288 (selinon) & II, p. 234

Silex
Salix L. ssp. esp. *viminalis* L. — Willow, esp. Withy, Osier
anglice a wythy, gallice sauz(C20)

Simphoniaca
Hyoscyamus niger L. — Henbane
gallice geneler [corr. gelener], anglice hennebelle vel hennebane(A1)
geneler [corr. gelener], hennebelle vel hennebane(A3) gallice geliner,

anglice henebane(A4) hennebane(A11)
henbane(B1) gallice gelyner, anglice hennebane(B2) hennebane(B3)
anglice henbane(C6)
Lit. André I, p. 308 & II, p. 253

Sinaparia see **Sinapis**

Sinapion
?*Crocus sativus* L. — Saffron
safran(B1)

Silocaracta [xilocaracta]
Rosa rubiginosa L. — Eglantine
Rosa canina L. — Dog Rose
eglentere(C19) anglice heepen(C19)
Lit. Alph., p.194; N221

Sinapis [sinapium, sinaperia, sinaparia, cinapis]
Sinapis L. ssp. and *Brassica nigra* (L.) Koch — Mustard(seed)/ Charlock
gallice senevé, anglice mustard(A1) cenevey, mustard(A3) [s]enevey, mustard(A3) gallice senevé(A4)
mustard(B1) anglice senevysed(B2) gallice senevey, anglice mostardsed(B2) senevoye(B8)
mustard(C2) senvy or mustardsede(C4) anglice werlok, mustardseed(C6) anglice zenevey(C15) henp, carlok(C28) sinvy or mostardsede(C31)
Lit. André I, pp. 294f. & II, p. 240

Sinnula [?]
Malva verticillata var. *crispa* L. — 'Crisped Mallow'
malva crispa, smerewort(B1)
Lit. B III, 207

Sinochidos [corr. linochidos]
Mercurialis perennis L./ *annua* L. — Dog's Mercury/ 'Annual Mercury'
anglice smeredoke(C3)

Sinonum [sinonium, scinonium, synamum]
Sempervivum tectorum L. — Houseleek
Smyrnium olusatrum L. — Alexanders, Horse-Parsley
gallice jubarbe, anglice selfgrene(A1) jubarbe, selvegrene(A3)
wylde garlek(B1)

wild persil seed(C10) seynt Mary flour*(C11) anglice stanmarche(C12)
wild perceli seed(C21) todisfenel, crescit super muros(C39)
Lit. cf. André I, p. 295 (sinon) & II, p. 241; Alph., pp.169, 18 & 172

Sinphitum [simphitum, sinfition]
Symphytum officinale L. — Comfrey
gallice confery, anglice muche brisewort(A1) confery, muche brisewort(A3) columbine*(A3) [co]lumbyne*(A5)
gallice confirie, anglice brosewort(B2) conferie(B3)
gallice confire, anglice muchel brusewort(C18)
Lit. André I, pp. 308f & II, p. 253

Sintilla
var. umbelliferae: *Cicuta virosa* L. — Cowbane
Anthriscus sylvestris (L.) Hoffm. — Cow Parsley, Keck
Angelica sylvestris L. — Wild Angelica
Conium maculatum L. — Hemlock
Aegopodium podagraria L. — Goutweed, Bishop's Weed
wodewhistil, kekkis(C21)

Siriaca [sciriaca]
Althaea officinalis L. — Marsh Mallow
malwes(C10, C11, C21) malewes(C24)
Lit. cf. André II, p. 241

Sirnys [sirmis]
Betonica officinalis L. — Betony
betoyne(C10) betony(C24)
Lit. N213 (sinirtus); Alph., p.173 (smircus); N213

Sisaris
?*Pastinaca sativa* L. — Wild Parsnip
?*Sium sisarum* L. — Skirret
siserioys(B1)
Lit. cf. André I, p. 295 (siser) & II, p. 241

Siseleos
Laserpitium siler L. — Sermountain
Siler montanum Crantz — Sermountain
Seseli montanum L. — Sermountain
gallice siler moyntayne(A1) sirmonteyne(A3)
gallice sermountaigne, anglice brotherwort(B2)
anglice broth[er]wort(C16) anglice brotherwort(C18) surmontayn(C22)

siler montayne(C21) sourmounteign(C24)
Lit. N214

Sisimbrium [zisimbrium]
Mentha sylvestris L. Horsemint
Mentha aquatica L. 'Water Mint'
gallice mentastre, anglice horsminte(A1) mentastre, horsmyntis vel watermintis(A3) anglice waterminte(A4)
anglice horsmynte(B2) horsmynte(B3) brokmente(B3)
brokmynte(C2) watermynte(C3) everfern*(C4) bawme*(C10) anglice groundeswel*(C15) horsmynt, spryngwort(C21) bawme*(C21)
Lit. André I, pp. 295f. & II, pp. 241f.

Sista
?*Capsella bursa-pastoris* (L.) Medic. ? Shepherd's Purse
staunche(C19)

Sistra
Anethum graveolens L. Dill
gallice ayné(C8) dille(C10) dill(C24)
Lit. cf. André I, p. 295 (sister) & II, p. 241; Alph., p.174

Sisula
Rhamnus catharticus L. Buckthorn
thethorn(C16)

Sitrunium
?*Eryngium maritimum* L. ? Sea Holly
anglice se-holy(C18)

Solago Maior
?*Hyssopus officinalis* L. ? Hyssop
ysope(B1)
Lit. cf. André I, p. 296 & II, p. 242

Solata
?*Calendula officinalis* L. ? Pot Marigold
anglice goldes(C29)
Lit. cf. André I, p. 297 & II, p. 243

Solatrum
Solanum nigrum L. 'Black Nightshade', 'Petty Morel'

Atropa bella-donna L. Deadly Nightshade, 'Great Morel'
?*Veratrum album* L. ? White Hellebore
morele(A7)
morele(B5)
(maius) more morel, madwort, (minus) houndeberye, houndebarie(C1) (nigrum) the more morell, (minus) þe lasse morele(C4) morell, mixber(C7) anglice nightshade(C18) anglice atterloþberie(C20) dwale, mekilwort, more morell(C21) (medium) medil morell(C21) wilde morele(C23) saynt Marie berie(C35) (medium, minus) houndysbane(C38) doworte, dowechs, lesior, more morell(C42)

Solonton
Potentilla reptans L. Cinquefoil
quinfoile(C21)

Solsequium
Calendula officinalis L. Pot Marigold
anglice rude(A2) solsekyle, roddis(A3) anglice rude(A6) anglice rude(A9) anglice golde(A12)
golde(B1) anglice rode flour(B2) rode(B3) gallice sousikyl, anglice marigold(B9)
(maius, domesticum) rodys (minus, agreste) wyld rude(C1) anglice marygoldys(C3) solsecull, goldes or ruddes(C4) goldwort, rede, holygold, rodewwrt oþer goldis(C7) golde, rode, giggeflour(C8) solsecle, marigoold, rudde, hardhoue(C21) solsecle, rodewort(C22) rodewort, solsecul, hardhowe(C32)

Sparagos Rusticos
Anthriscus sylvestris (L.) Hoffm. Wild Chervil
woodphille(C21)

Sparagus [spargus, speragus]
Geranium molle L./ *columbinum* L. 'Dove's-Foot Cranesbill'/ 'Long-Stalked Cranesbill'
?*Ruscus aculeatus* L. ? Butcher's Broom
sparge(B1) gallice sparage(B2) culverfot(B4) sparge, colverfot(B8)
(medius) mawewort(C4), sparge(C1) colverevot(C2) anglice brusche(C3) (speragus) mawort(C4) anglice spurge*, et secundum alios cullerage*(C6) coloverwot(C7) colverefot(C12) gallice sperage, anglice colverfot(C18) sparge, mawort(C21) bruscus, est frutex spinosa, anglice bruske(C29) (medius) mawwort(C38)
Lit. Agnus Castus pp. 247f; Alph., p.168, 6 (speragus)

Spargana [sparganus]
Sparganium L. ssp. ? Bur-reed
fouþistel(B5)
(maior) spurge(C1) wodethistel(C8)
Lit. cf. André I, p. 298 (sparganion) & II, pp. 243f.

Spargia
?*Asparagus officinalis* L. ? Asparagus
?*Geranium molle* L./ *columbinum* L. ? 'Dove's-Foot Cranesbill'/ 'Long-Stalked Cranesbill'
sparge(B1)

Spatula Fetida
?*Iris foetidissima* L. ? Gladden, Stinking Iris
gladyn þe which beryth no flowyr(C11) anglice gladyn(C16) gladene(C18) gladiol sine flore(C21) it haþ no flour(C24) gladyn bering no flour(C24)
Lit. Alph., p.164, 6

Spelta
Triticum spelta L. Spelt
anglice splet(C1)
Lit. André I, pp. 298f & II, p. 244

Spergula [spargula, spargulum, spertula, spagula]
Galium cruciata (L.) Scop. Crosswort, Mugwort
?*Spergula arvensis* L. ? Corn Spurrey
?*Geranium* L. ssp. esp. *molle* L. ? (Dove's-Foot) Cranesbill
(minor) anglice crowser(A2)
marchewort, spurge*(B4)
(maior) crowworte (media) cruse (minor) herbe paule, croswort(C1) (minor) secundum aliquos culrage*(C5) spurge*(C7) (maior) croyse(C8) herbe croyce(C13) (maior) mor croswort(C21) (maior) cruciata, S. Mary wort(C32) the more and the lesse hath his levys departid as fendesbyte(C33, C37) (maior) crowwort (minor) croswort(C38)
Lit. Alph., p. 79, 4 & 168,1; A9 = 'spergula maior, herba cruciata idem; stipitem erectum habet, folia circumquaque gradatim, minores habet flores superius croceos'; Agnus Castus pp. 247f.

Sperula
Ranunculus flammula L. Lesser Spearwort
sperewort(A3) anglice sperewort(A5)

Spica Celtica
Valeriana celtica L. Celtic Spikenard
?*Ophioglossum vulgatum* L. Adder's Tongue
gallice spike seltic(B2) neddre-spere(B3)
spykenarde(C11) spikenard(C21) morsulk(C23) anglice cattstayle or calketrape or cawetrapp(C26)
Lit. André I, p. 299 & II, p. 245

Spicarius
?*Geranium columbinum* L. ? 'Long-Stalked Cranesbill'
maworte(C42)
Lit. Agnus Castus pp. 247f.

Spigurnella [spurginelle]
?*Meum athamanticum* Jacq. ? Spignel, Meu
?*Silene dioica* (L.) Clairv./ *alba* (Mill.) E.H.L. Krause/ *noctiflora* L. ? Red- /White- / 'Night-flowering' Campion
spigurnelle(C21) .iii. species sunt .i. spurginel(C32) spigurnall (rubea) red spigurnell(C38) anglice spigurnell [3 varieties: white, red and yellow flowers]
Lit. Alph., p.174; Agnus Castus p. 245; C32 = bardona

Spillum [for serpillum?]
Mentha pulegium L. Penny-Royal
Thymus serpyllum L. Wild Thyme
gallice puliol, anglice hilwort(C16)

Spina Alba
Crataegus monogyna Jacq. Hawthorn
?*Berberis vulgaris* L. ? Barberry
anglice whit thorn(C16) hawthorn(C36)
Lit. André II, pp. 245f; N215

Spinacia
?*Spinacia oleracea* L. ? Spinach
gallice spinache, anglice spinogre(B2) anglice spinache(B7)
spynache(C1) spynnache(C4) gallice spinache, anglice spinegre(C18)

Spinagra
?
spinogre(C23)
Lit. Alph., p.174 (spigurnella)

Spinarchia
Spinacia oleracea L. Spinach
anglice spinarch(C12) caulus non plantatus, anglice spynarche(C26)
Lit. cf. Agnus Castus p. 247

Sponsa Solis
Lithospermum officinale L. Gromwell
Calendula officinalis L. Pot Marigold
Cichorium intybus L. Chicory, Wild Succory
gallice gromyl(A1) gromoyl(A3)
gallice et anglice gromyl(B2) cicori(B3)
anglice gromyll(C6)
Lit. André I, p. 301 & II, p. 247

Spurgia
Euphorbia L. ssp. Spurge
anglice clensynggras vel spurge(C16)

Squilla
Scilla L. esp. *Scilla maritima* Baker Squill/ Sea Squill
squill(A3)
un maner de cepe(B2)
(minor) sanylle [?](C1) anglice see-oynyon(C4) se-onyon(C21)

Squinantum [squinacium]
Andropogon schoenanthus L. Camel's Hay, Squinant
squinant(A3)
gallice et anglice squinaunce(B7)
gallice et anglice squinant vel sclavelor(C5) camelismete(C10)
squynant(C22) camelismete(C24) anglice camylsheye(C26)
Lit. André I, p. 284 (schoenanthus) & II, p. 229; Alph., p.174

Stacados [for sticados?]
?*Sempervivum tectorum* L. ? Houseleek
gallice jubarbe, anglice sengren vel houslek(B1)
Lit. Alph., p.175; N215

Staphisagria [stasisagria]
Delphinium staphisagria L. Stavesacre
staphisagre(A3)
staphisagre .i. schabbewort(B1) anglice stafizagre(B2) wilde byne [corr. vyne](B4) anglice stasisagre[sic](B7)
stafisagur, wild vyne(C4) anglice stasisagre [sic](C6) herbe pur netis and

lyes(C10) herba pro nyttes and lyes(C11) anglice lous-erbe(C29) lausword(C32)
Lit. André I, p. 302 & II, p. 248. The plant, also known as 'herba pedicularis/ pediculosa' and 'uva agrestis', was used in the treatment of lice, see Agnus Castus p. 247

Starongaria [= scaron + agria?]
?*Origanum vulgare* L. — ? Marjoram
?*Mentha pulegium* L. — ? Pennyroyal
?*Thymus serpyllum* L. — ? Thyme
organ(C21)

Stellaria
?*Alchemilla vulgaris* L. — ? Lady's Mantle
?*Aster amellus* L. — ? Aster
?*Plantago coronopus* L. — ? Buck's-Horn Plantain
(maior) more sterwort(C1) sterrewort(C4) sterwort(C21) (maior) sterrewort (minor) lasse sterrewort(C32) (minor) lasse starwort(C41)

Sterobileon
Pinus pinea L. — Pine-Nut, Pine-Cone
pinee(B3)
Lit. André I, p. 305 (strobilus); N216

Sticados [citrinum]
Sempervivum tectorum L. — Houseleek
citrin(A3)
citryne(C22) housleke(C35)
Lit. Alph., p.175; N215

Sticidiagria [sticidiagna]
Ecballium elaterium A. Rich. — Squirting Cucumber
anglice gourden(B2)
gallice [sic] gurden(C16) anglice gourden(C18)

Stinncem [?]
?*Arctium lappa* L. — ? Burdock
burre, clete(C21)

Stinum
Silene alba (Mill.) E.H.L. Krause — White Campion
Silene vulgaris (Moench) Garcke — Bladder Campion

cowratil(C21)
Lit. Agnus Castus p. 219

Stipteras [?]
?*Lavandula officinalis* L. — ? Lavender
gallice lavendre(B1)

Strignum [stringium, stringnum]
Solanum nigrum L. — 'Black Nightshade', 'Petty Morel'
gallice morelle, anglice atterloueberi(A1) morel, attirloueberie(A3)
grant morele(B3) morell(B7)
graunt morelle(C2) nightshede(C4) anglice morell(C6) nytisshede, ilhaw, litil morel(C21) morel, arterloþeberie(C22)
Lit. André I, pp. 305f & II, p. 251

Strisosilla
Centaurium erythraea Rafn — Common Centaury
lesse centory, cristeleddre(C21)

Strutium [strution, osturcium, ostrucium, sturcium, osturse]
Brassica oleracea L. — Wild Cabbage
Saponaria officinalis L. — Soapwort
Peucedanum ostruthium (L.) Koch — Master-Wort
Crambe maritima L. — Sea-Kale
anglice wylde coule(A1) wylde coul(A3) bissophiswort(A3)
bychoppewort(A6) biscopiswort(A9)
anglice byschopeworte(C3) alysaunder(C7) smale caule(C10) smal caul(C11) cockles or wild cale(C14) elilaundre [corr. elisaundre](C19)
anglice bissopiswort(C20) smale caule(C21) wild wort(C30)
cherlok(C31)
Lit. N216; Macer p. 221; see **Osturce**

Strutium 2 [error for strignum?]
Atropa bella-donna L. — Deadly Nightshade, 'Great Morel'
Solanum nigrum L. — 'Black Nightshade', 'Petty Morel'
morel or houndbery(C21)

Sucofulon
Althaea officinalis L. — Marsh Mallow

holihok, wild malwe(C21)
Lit. André I, p. 308 (sycophyllon) & II, p. 253

Suffonia
Helleborus niger L. — Black Hellebore
wodeþung(A3) wodeyu[n]g(A5)

Sumac
Rhus coriaria L. — European Sumac
gallice et anglice asimac(B2)

Supercilium Veneris
Achillea millefolium L. — Yarrow, Milfoil
nozebledels(B3)
anglice ȝarwe(C10) anglice ȝarowe(C16) yarwe, millefoile(C21)
Lit. André I, p. 307 & II, p. 252

Suphitum
conf. of Alum (= symphytum) with Allium
ayl, garlek(B1)

Talmon [for buphthalmon]
Chrysanthemum leucanthemum L. — Ox-Eye Daisy
Primula veris L. — Cowslip
kowslep(C4)

Talpis
?*Mercurialis perennis* L./ *annua* L. — ? Dog's Mercury/ 'Annual Mercury'
mercurial(C13)
Lit. Alph., p.181 (thalphi)

Tamariscus [tamaruscus]
Tamarix gallica L. — Tamarisk
?*Phoenix dactylifera* L. — ? Date Palm
anglice date(A6) anglice date vel gallice(A8) anglice date(A9, A12)
date(B3)
ivy*(C8) moleyne*(C10) yvy*(C13) anglice date(C15) tamaryndes* or brome*(C21) anglice quycbeme(C26) brome*(C29)

Tanacetum [tanaceta, tansetum, tanesetum]
Chrysanthemum vulgare (L.) Bernh. — Tansy
Potentilla anserina L. — Silverweed

taneseye(B1) tanasi(B3) tansey(B4) taneseie(B5) (domesticum) anglice tanesy(B9)
tansy(C1) gosegras(C2) anglice gosegrys(C3) anglice tansey(C6) fedirfew(C21) (ortulanum) tansey(C29)
Lit. André I, p. 311 & II, pp. 255f.

Tanacetum Agreste [tancetum a., tanesetum a.]
Potentilla anserina L. Silverweed
anglice gosegres(A2, A12) gosegres(A6) anglice gosegras(A8) anglice gosegresse(A9)
gosegres(B2) gosegras(B3) gallice tanesé, anglice gosegresse(B9)
anglice gosegres(C6) wild tanzay, gosgrasse(C13) wild sauge*(C29)

Tanasia
Chrysanthemum vulgare (L.) Bernh. Tansy
Potentilla anserina L. Silverweed
(maior) tansy (minor) wyld tansy(C1) (campestris) wilde tansy, gosegrasse, eglamour(C4)
Lit. Alph., p.181 & **Athanasia**

Tapsia
Thapsia garganica L. 'Deadly Carrot'
conf. w. Tapsus barbastus 'Mullein'
gallice tapsie(A1) velwort*(A3)
anglice wodenote(B1)
gallice tapsye(C6) softwort*(C7) gallice tassie(C18) wild sporge*(C21) faytours[wede](C37) faytours herbe(C41)
Lit. Alph., p.182; André I, p. 313 (thapsia) & II, p. 313

Tapsus Barbastus [tapsis b., t. barbatius, t. barbariscus]
Verbascum thapsus L. Mullein
gallice moley[n](A2) moylene, veltewort(A3) gallice moleyne, anglice feltwrt(A4) gallice moleyne(A6, A12) moleyne(A7) gallice molayne(A8, A9)
moleyne(B3) molayne(B4) henebane*(B5) molene(B5)
moleyne, feldwort(C1) molyn(C3) [l]ungwort, feltwort, moleynwort(C4) walewort*, moleyne, foxeye(C8) veltwrt, moleyne, foxoȝȝe(C13) longewort, tasyl(C30) moleynis, feltwort, hegge-tapir(C31) soft, moleyne(C33) wild moleyn(C35) softe, moleyn(C37)
Lit. Alph., p.182; N217

Tarascon [for taraxacon]
Sonchus oleraceus L. Milk- or Sow-Thistle

Lactuca serriola L. — 'Prickly Lettuce'
scariole(C10, C11, C21, C24)

Temolus [temelus]
Sempervivum tectorum L. — Houseleek
zyngrene(B3)
anglice singrene(C12)
Lit. B III, 208

Tenaceta
?*Pistacia terebinthus* L. — ? Terebinth, Turpentine
terbentyne(C21)

Tenericium [?]
?*Pirola rotundifolia* L. — ? 'Winter-Green'
pirole(B1)

Teorioston
?
hunisocle(C21)

Terebentina
Pistacia terebinthus L. — Terebinth, Turpentine
gallice terebintine(A1) terebentine vel turbentyn(A3)
gallice et anglice terbentine(B2)

Terentis
?*Thapsia garganica* L. — ? 'Deadly Carrot'
gallice tassie(B2)

Terraneus [?]
?*Convallaria majalis* L. — ? Lily-of-the-Valley
glofwort(B3)

Terreioa
?*Phyllitis scolopendrium* (L.) Newm. — Hart's-Tongue Fern
hertistong, cerlang(C21)

Tesapium
Sinapis alba L. — White Mustard
gallice blanc moustard(B2)
blanc mustard(C2, C18) gallice blank mustard(C16)
Lit. Alph., p.184; André I, p. 315 (thlaspi) & II, p. 259; N218

Testiculus Auris [for muris]
?*Sanguisorba officinalis* L. ? 'Great Burnet'
?*Anagallis arvensis* L. ? Scarlet Pimpernel
selfhele, waywort, morecrop(C21)

Testiculus Leporis
Orchis L. esp. *Anacamptis pyramidalis* (L.) Rich. Orchis/ 'Pyramidal Orchid'
'satureia rubea'?
ravenleke(C21)
Lit. André I, p. 312 & II, p. 257

Testiculus Muris
conf. w. auricula muris
?*Potentilla anserina* L. ? Silverweed, 'Goose Tansy'
mowseheer*(C37) gosgrece(C39)
Lit. Alph., p.184

Tetrahit [tretraid]
Tragopogon pratensis L. Goat's-Beard
Galium aparine L. Cleavers
clyvere(C2) clive(C7) saliunca, anglice kaltrappe*(C19)

Tharmos
Lupinus L. ssp. Lupin
lupynes(C21)
Lit. André I, p. 314 (thermos) & II, p. 259

Thesum
Mentha piperita L. Peppermint
white mynt(C21)
Lit. André I, p. 312 (tesmu)

Thymus
Thymus serpyllum L. Thyme
anglice time(A4)
tyme(B1)
tyme(C1)

Tiarisa
?*Betonica officinalis* L. ? Betony
betoyn, bisshopwort(C21)

Tiaspus [for thlaspis?]
?*Capsella bursa-pastoris* L. Medic. ? Shepherd's Purse
?*Thlaspi arvense* L. ? 'Field Penny-Cress'
lombeshert(B1)
Lit. B III, 154f. cf. André I, p. 315 (thlaspi) & II, p. 259

Tilia
Tilia L. ssp. Linden
lynde(C6)

Timbra [timbria, tumba]
Satureia hortensis L. Summer Savory
Satureia montana L. Winter Savory
sathiré(A3)
gallice et anglice saþereie(B2) (tymbria) tymbre(B3)
saveray(C3) gallice sacher(C6) savere(C7) savereyne(C11) sacher(C13)
saverey(C21) bisworte[?](C24)
Lit. André I, p. 315 & II, p. 260; Alph., p.186

Tiriaca [t. rusticorum]
Allium sativum L. Garlic
?*Allium porrum* L. ? Leek
leke(C14) garleke(C21, C26)
Lit. Alph., p.187

Titimallus [tintimallum, titimalli, titemallus]
Euphorbia L. ssp. Spurge
Sempervivum tectorum L. Houseleek
spurge(A4, A7)
marceure, singrene(B1) spu[r]ge, glotiner*, clote*(B8)
tytymalle, wyld spurge(C1) anglice watergrys [for wartgrass?](C3)
plures sunt species earundem, walworte, esula, anabula, catapucia,
laureola(C24) = þer be many spics of hem, wallewort, esula, anabulla,
catapucia, laureola and other(C11)
Lit. B III, 208; Alph., p.185; André I, pp. 317f. & II, pp. 261f.

Tora [thora]
Ipomœa turpethum L. Jalap
gallice et anglice turbith(B2)
turbyth(C13) gallice et anglice turbit(C16) spereworte*(C29)

Tormentilla [turmentina]
Potentilla erecta (L.) Räusch. Common Tormentil

gallice tormentine(A1) tormentille vel tormentine(A3) gallice tormentine(A4)
gallice tormentine, anglice turmentille(B2) tormentille(B3) turmentill(B4) hote note, treakyl dé boys(B4) turmentine(B8)
turmentylle(C1) cornеþuohe or hoytnote(C2) anglice tormentyll(C3) tormentilla [sic], kympilmore(C21) romani vocant turmentyne, anglici hetenote(C43)

Torpedelopion [tarpidolopium, tarpolodium]
Plantago major L. 'Great Plantain'
playnteyn, weybred(C21)
Lit. cf. André I, p. 311 (tardibolotius)

Toxiobulus [toxobulus]
Artemisia vulgaris L. Mugwort
mugwort(C21)
Lit. André I, p. 318 (toxobolus) & II, p. 262

Tremulus
Populus tremula L. Aspen
anglice asp(B2)
aspe(C6) anglice aspe-tre(C26)

Tribulus
Rubus L. ssp. Bramble
?*Rosa canina* L. ? Dog Rose
?*Ulex europaeus* L. ? Gorse
gallice runce, anglice brembel(B2) gorst(B3) brere(B5)
gallice rounce, anglice brer(C6) rounce, reed brere(C10) reed brere(C21)
Lit. B III, 115

Tridax
Lactuca virosa L./*serriola* L. Lettuce/ 'Prickly Lettuce'
wild letuse(C21)
Lit. André I, p. 315 (t(h)ridax) & II, p. 260

Trifolium [trifolion]
Trifolium L. ssp. Clover, Trefoil
gallice trifoyle, anglice þre-levede gras(A1) trifoil, clavergras, þre-levyde gras(A3) þre-levyde gras, samrok(A3)
gallice trefoil, anglice þre-lef(B2) clovere(B5)
anglice honysokes, thre-levyd gras(C4) anglice þre-levyd clover(C6)

trefoylgres(C29) . . . anoþir þat grouiþ comynly among whete and ys calid corona regia, kyngys-crowne(C31)

Trimomanicos
?*Atropa bella-donna* L. ? Deadly Nightshade, 'Great Morel'
?*Datura stramonium* L. ? Thorn-Apple
foxglove(C21)
Lit. André I, p. 306 (strychnon manicon) & II, p. 251; B III, 99

Trisago Maior
Teucrium chamaedrys L. 'Wall Germander'
germandre(C21)
Lit. André I, p. 322 (trixago) & II, p. 265

Trisogonos
Teucrium L. ssp. Germander
germandre(C21)
Lit. cf. André I, p. 88 (chrysogonos) & II, p. 64

Troxima [toxia, toxima]
Sonchus oleraceus L. Milk- or Sow-Thistle
Lactuca serriola L. 'Prickly Lettuce'
Cichorium intybus L. Chicory, Wild Succory
endive, scariole(C10) horsthistil, endive(C21) endive, scariol(C21)
Lit. André I, p. 322 & II, p. 265

Truner [trumer, tumor, tuner]
Nymphaea L./ *Nuphar* Sm. ssp. Water-Lily
anglice waturlilie(B7)
anglice waturlyly(C6) edok, waterlilie(C8) edok, waturlily(C13) edoc(C23)

Truta Minor [trutti m.]
Thapsia garganica L. 'Deadly Carrot'
anglice faytirwort(C4) faytorswede, fatorswort(C21)
Lit. see **Herba Trutannorum**

Tuber
var. *agaricaceae* Mushroom, Toadstool, Fungus
paddok-chese(C28)

Tubera
Cyclamen hederifolium Ait. Sowbread
Conopodium majus (Gouan) Loret Pignut, Earthnut
anglice erthnote(B1)
Lit. André I, p. 322 & II, p. 266

Tucilla Rustica
Origanum dictamnus L. Dittany of Crete
Dictamnus albus L. White Dittany
ditony(C21)
Lit. André I, p. 323 (tussella rustica) & II, p. 266

Turbit
?
anglice turmeryc(C12)

Turiones
Vitis L. Vine-Shoots
þe croppys of vynes, whit or reed(C10) croppes of vyne(C21)
Lit. André I, p. 323

Tuxitemus
?*Aristolochia rotunda* L. ? Aristolochia
?*Cyperus longus* L. ? Galingale
moke, galingale(C21)

Uligo
var. algae
flotgres(B3)
flotgras(C1) flo[t]grasse(C2) anglice flodgrasse(C12)

Ulmus
Ulmus L. ssp. Elm
anglice elme(A10)
elm(C2) omel, elm(C8) anglice peniwort*, gallice omel, anglice elme(C12)

Umbilicus Veneris [umblicus v., umbilicum v., embelicus v.]
Umbilicus rupestris (Salisb.) Dandy Pennywort
?*Cymbalaria muralis* Gaertn., Mey. & Scherb. ? Ivy-Leaved Toadflax
deneroil, peniwort(A3) peniwort(A7) peniwrt(A8)
peniwort(B1, B3) peniwrt(B5) anglice penigres(B9)

penywort(C1) anglice penygrysse(C3) anglice peniwort or wild tanseye*(C12)
Lit. André I, p. 335 & II, p. 275

Unctuosa
?*Mercurialis perennis* L./ *annua* L. — ? Dog's Mercury/ 'Annual Mercury'
'Selfheal' & var. unidentified plants, poss. *Aristolochia rotunda* L.
uncterele(A8) gallice uncturol(A9) gallice uncturole(A12)
oynterole(B3) gallice uintur[o]le(B9)
(minor) smerwort(C1) smerwort(C4) gallice uncturiol, idem est quod ratyl(C6) uncterol, ratil(C8) uncterole, ratyl(C13) (u. minor) smerwort(C21) unterole, anglice rathele, smerewrt(C23)
Lit. Alph., p.193 (notes); see **Ointerrele**

Ungula Caballina
?*Asarum europaeum* L. — ? Asarabacca
?*Tussilago farfara* L. — ? Coltsfoot
(terrestris) pappwort, enny lyttel clote(C6) crowgold, papwort(C7) papwort(C12) (terrestris et aquatica) anglice horshove(C29)
Lit. Alph., p.193; SB 43; André I, p. 335 & II, p. 275

Ungula Caballina Aquatica
Nymphaea L./*Nuphar* Sm. ssp. — Water-Lily
nenufar, waterlylie(A3)
edokke(B3)
water-rose(C3) anglice waturlyly(C6) þe rote of waterlilie or elles of menphie(C10) the rote of waterlyly or of memphe(C21)
Lit. Alph., p.192; André I, p. 335 & II, p. 275

Unturola
?
anglice moderwort(B1)
Lit. cf. **Ointerele**

Upia
?*Cinnamomum cassia* (L.) Blume — ? Cinnamon
canel(C8, C13)

Urtica
Lamium L./*Urtica* L. ssp. — Nettle
netle, urti(A3)

urteye, nettyll(B8)
(alba) whyt netyl (rubea) rede netyl(C1)

Urtica Alba
Lamium album L. White Dead-Nettle
ded-netle(C35)

Urtica Ceca
Lamium album L. White Dead-Nettle
blynde netell(C7)

Urtica Greca
?*Urtica urens* L. ? Small Nettle
urtie, netle(A3)
gallice rouge urtie, anglice rede netle(B2)
netyl(C1) grekys nettill, rede nettyll(C4) rouge urteie, reed netel(C10) gallice rouge urti(c)e, anglice red netil et blyndnetyll(C16) white nettyll(C21) reed nettill(C21) wilde nettill(C24) crekis-netle(C35)

Urtica Mortua
Lamium album L. White Dead-Nettle
Lamium purpureum L. Red Dead-Nettle
(alba) whyt dede-netyl (rubea) red ded-netyl(C1) blynd netel, wild netel or deed-netel(C10) dede-nettill, archaungell(C21) wylde netel(C32)

Urtica Rubea
Lamium purpureum L. Red Dead-Nettle
anglice scherp-nettill(C29) red netle(C35)

Urticella [for viticella]
Bryonia dioica Jacq. Bryony
smernepe(C1) bryonye(C37)
Lit. see **Vitiscella**

Urticolla
?
wodebind(B1)
Lit. see **Vincella**

Uspila
?
anglice ersope(C5)
Lit. cf. MED ere-sope (= earwax)

Uva
Vitis vinifera L. Grape
grape(A3)
grape(B2)

Uva Lupina
Atropa bella-donna L. Deadly Nightshade, 'Great Morel'
Solanum nigrum L. 'Black Nightshade', 'Petty Morel'
morele(B3) morell, nythsode(B4)
pappewort*(C4) morell, nighcode(C7) houndesberie(C15) morell, houndbery(C21) pety morel(C31)
Lit. André I, pp. 337f & II, p. 277

Uva Passa
Vitis vinifera L. Grape
reysins(A3)

Vaccinium [vaccinum]
Vaccinium myrtillus L. Bilberry
Primula veris L. Cowslip
blakeberyes(C2) viola nigra, anglice cowesloppys(C3) kowslop(C4) anglice blacberyon vel mulberion(C6) bulleberry(C6) mulberye(C8) hoorten(C10) hurtenem(C11) blakbery, mulbery(C13) vaccineus, herba paralesis, couslop(C13) mirtilles, hurtene(C24)
Lit. B III, 141 & 41; André I, p. 325 & II, p. 268

Vaginella
Rosa Canina L. Hip
hoppys(C4) heppes(C7)

Valeriana
Valeriana officinalis L. Valerian
valeriane(A3)
valerien(B1) valerian(B3)
walerian, lovewort, lowewort(C1) anglice valeriane(C3) valerian, hertclyver(C4) zeduare agreste, anglice maiseles*(C5) zeduare agreste, anglice maseleys*(C6) valerian, hert[c]lyver, lovewort(C21) valerian, lovewort, hertewort(C32)

Valerianus
?Sonchus oleraceus L. ? Milk- or Sow-Thistle
sowthistel(C15)

Veneria
Rubia tinctorum L. Madder
maddir(B1)
Lit. cf. André I, p. 326 & II, p. 268; see B III, 167

Venter Apis
Achillea millefolium L. Yarrow, Milfoil
gallice milefoyle, anglice yarow(A1) millefoyl, yarou(A3)
gallice milfoil, anglice ȝaruwe(B2)
ȝarwe(C10) yarwe(C21)
Lit. Alph., p.190 (note) & N193 (sub millefolium) & N220

Venti Med[i]a [i.e. herba v.m.]
Ajuga reptans L. Bugle
bugle(B1)
Lit. Alph., p.190

Veprium Folia
Rubus L. ssp. Bramble-Leaves
bumble-lef [corr. bramble-lef?](B2)
bremil-levis(C1) anglice hemyl-leves [sic](C16)

Verbalis Maior
Agrimonia eupatoria L. Agrimony
egrimony, coseclite(C21)
Lit. André I, p. 326 & II, p. 269

Verbascum [verbasca]
Verbascum thapsus L. Mullein, Aaron's Rod
moleyn, feldwort(B1) feldwort(B3)
anglice feltwort(C12)
Lit. André I, pp.326f. & II, p.269

Verbena
Verbena officinalis L. Vervain
gallice verveyn(A2) gallice verveyne(A12)
verveyne(B3)
werveyne(C1)

Vercardonum
?Cnicus benedictus L. ? Holy Thistle
thowthistill(C21)

Vermicularis
Sedum acre L. Common Stonecrop
test de soriz, anglice stoncrop(B1) anglice stonore(B2) stoncrop(B3) anglice staynchorppe(B9)
anglice stonecroppe(C3) anglice stonore(C4) teste de sorre, anglice stancrop vel ventolayne(C5) gallice teste de sar[is], anglice stancrop vel mecalayn [sic](C6) stanore, o[r]pyn, orvale(C7) anglice stonhore(C12) (maior) orpyn, (minor) stoncroppe(C14)
Lit. André I, p. 327 & II, p. 270

Vermicularis Maior
Sedum telephium L. Orpine
orpin(A2) orpin(A3)
orpin(B1)
Lit. Alph., p.190

Vermicularis Minor
Sedum acre L. Common Stonecrop
Sedum telephium L. Orpine
peniwort(A2) peniwort(A3)
teste de soriz(B1)
Lit. Alph., p.190

Verminacia [verminaca]
Sedum telephium L. Orpine
orpyne(C14) dow(n)wort*, columbyne*, cokfote*(C21)
Lit. cf. André I, p. 328 & II, p. 270

Verrucaria [verucoria]
Calendula officinalis L. Pot Marigold
Cichorium intybus L. Chicory, Wild Succory
gallice cycoré(A2) cicorie(A3)
solcicle, anglice golde(B1) rode(B3)
roodes(C10) anglice roden(C12)
Lit. André I, p. 328 & II, p. 270

Vervena
Verbena officinalis L. Vervain
verveyn(A3) gallice verveyne(A6) gallice vervayne(A9)

verveyne, anglice ylenhard(B1) vervayne(B4, B9) verveyne(B5)
anglice vervayne(C3)

Vessica Campi
?*Silene vulgaris* (Moench) Garcke ? Bladder Campion
feld bleddyr(C38)

Vetonica [betonica, betonia]
Betonica officinalis L. Betony
betoyne(A3)
betoyne(B3) bisschopwort(B5) vetoyn(B6)
idem quod febrifuga, feverfoy(C4) fetherfoye(C7)
Lit. André I, p. 329 & II, p. 271

Vetrum [for veratrum]
Veratrum album L. White Hellebore
lungwort, elebre(C21)

Vibex
Betula pubescens Ehrh. Birch
Betula pendula Roth Silver Birch
anglice berke(C26)

Vicentia
Achillea millefolium L. Yarrow, Milfoil
millefoile, yarwe, nosebledel, nesegresse(C21)
Lit. André I, p. 329 (vigentia)

Vicetoxicon [vicetoxicum]
?*Gentiana amarella* (L.) Börner ? Felwort
?*Mercurialis perennis* L./ *annua* L. ? Dog's Mercury/ 'Annual Mercury'
?*Potentilla erecta* (L.) Räusch. ? Common Tormentil
anglice baldemoyne(C19) mercuri, papwort(C21) mercurie(C33) anglice tormentill(C36)
Lit. cf. N220

Victorala
?*Bellis perennis* L. ? Daisy
?*Viola* L. ? Violet
bonwort(B1)
Lit. cf. André I, p. 329 (victoriola) & B III, 16f (banwyrt)

Vigellum
Agrostemma githago L. — Corn Cockle
Lolium temulentum L. — Darnel
pobel(A3)
pobell(B7)
poble(C5) pebel(C6) pobell(C21) pobyl(C22)

Vimen
Salix viminalis L. — Osier
oseer(C13)

Vinca
Vinca major L./*minor* L. — Greater/Lesser Periwinkle
pervenke(B3) pervenk(B5)

Vincella [for viticella?]
?*Lonicera periclymenum* L. — ? Honeysuckle
?*Convolvulus arvensis* L. — ? Bindweed
wodbynd(C3)
Lit. cf. B III, 264f.

Vinicium
?
foleslek(B1)

Viola
Viola L. ssp. — Violet
gallice violes, anglice violet(A1) violete(A3) gallice violet(A4)
gallice et anglice violet(B2) her[t]worte, violette(B4)
vyolet(C1) (maior) hermole, violet(C4) anglice hertwort(C12) comon violet(C21) romani vocant eam violet, anglici similiter vel slewort(C43)

Viola Agrestis/Subalbida
?*Viola canina* L. — ? 'Heath Violet'
nay[l]wort, ungylwort, lasse wyolet(C1)
Lit. see **Viola Minor**

Viola Alba
?*Viola* L. ssp. — ? Violet
?*Hesperis matronalis* L. — ? Dame's Violet
gallice raycele(B2)
anglice racyl(C6) galice raicell(C16) gallice raycele(C18)

cowslope*(C21) anglice white vyolet(C26) anglice ratyl(C36)
Lit. Alph., p.191; N220

Viola Aurosa
?*Viola* L. ssp. ? Violet
?*Cheiranthus cheiri* L. ? Wallflower
hennewort(B1)
Lit. Alph., p.191

Viola Minor
?*Viola canina* L. ? 'Heath Violet'
anglice þe lasse violet, ungulwort or maylewort [corr. naylewort](C4)
ungwort [sic], nailwort(C21) ungulwort, nayl(C32)

Viole
?
viole, anglice stubwort(B1)

Viperina [herba v.]
Polygonum bistorta L. Bistort, Adderwort
Dracunculus vulgaris Schott Dragon Arum
serpentine, dragaunz(B1)
dragance(C21) naderwort(C37)
Lit. André I, p. 331 & II, p. 273

Virga Pastoris [v. pastorum]
Dipsacus fullonum L. Teasel
anglice wilde tesle(A2) wylde þistel(A3) anglice wylde tesele(A6) anglice wilde tesel(A8) wilde tasel(A9) anglice wilde tasele(A12) anglice wilde taseles(B2) wilde tesil(B3) wylde tasell(B4) wyld tasyll(B8) anglice wild tasel(B9)
wyld tesyl(C1) wyld tasyll(C3) wilde tesil, alomoth(C8) wilde tasel, alomoth(C13) anglice wilde tasel(C20) wilde thistell(C22) scheperdes yerd(C41)

Virginalis
Mercurialis perennis L./*annua* L. Dog's Mercury/ 'Annual Mercury'
smerewort(B3)
unctuosa idem, smerwort(C1)
Lit. André I, p. 331 & II, p. 273

Viscago
Ilex aquifolium L. Holly
hulver, þe grete holyn(C9)

Viscerago
Viscum album L. Misteltoe
uolvescombe* [= virga pastoris](B1) mistilto(B8)
mistilto(C6)
Lit. André I, p. 332 (viscarago) & II, p. 273

Viscus
Althaea officinalis L. Marsh Mallow
?*Polypodium vulgare* L. ? Polypody
wymawe(A3)
anglice wilde fern(C12)
Lit. André I, p. 332

Viscus Quercinus [v. querci]
Viscum album L. Mistletoe
wy de chene(A3) wy de chene(A5)
uie de cheine(B1) wy(l) de chene(B4)
mystel of þe ooke(C1) mistell of an ok, wy(the) de cheyne,
misteltayn(C2) anglice mystill of þe oke(C4) uy de cheyn(C7) wi(l) de
keyn or mystelon of oke(C10) mistelto(C19) mystill(C21)
Lit. André I, p. 332; Alph., pp.191ff.

Vitis Agrestis
Vitis sylvestris Gmel. Wild Vine
gaser(A3)
gallice gasir(B2)
gallice [g]aser(C6) anglice gazer(C12) gallice gasir(C18) gazer(C22)

Vitis Nigra
Bryonia dioica Jacq. Red Bryony
?*Tamus communis* L. ? Black Bryony
bryan(C4) briane(C21)

Vitis Silvestris
Vitis sylvestris Gmel. Wild Vine
?*Tamus communis* L. ? Black Bryony
vin savage(A3)

Vitiscella [vitis corella, vitis cerella, vitis corolla]
Bryonia dioica Jacq. Red and White Bryony
anglice smernep(B2) gallice uve passe*, anglice reysins*(B2) anglice smerenop(B7)
anglice sermenep[sic](C6) anglice smernepe(C12) smernep(C13) smernepe(C21)
Lit. André I, p. 332; SB 43

Vitraria
Parietaria diffusa Mert. & Koch Pellitory-of-the-Wall
vitriola, litewort, hemerwort(C21)
Lit. André I, p. 333 & II, p. 274

Vitriola [vitriolis]
Parietaria diffusa Mert. & Koch Pellitory-of-the-Wall
paritarie(A3) paritarie(A7)
gallice vitriole(B2)
peritory(C4) gallice viturol(C6)
Lit. André I, p. 333

Volubilis
Lonicera periclymenum L. Honeysuckle
Convolvulus arvensis L./ *Calystegia sepium* (L.) R.Br. Bindweed/ 'Larger Bindweed'
Hedera helix L. Ivy
weþewynde(A3) wodebonde(A3)
anglice yvi(B2) wodebynde(B3) anglice wodebind(B9)
(maior) wodebynde, wythywynde(C1) anglice wodbynde(C3) (maior) wythwynde, (minor) wodebynde(C4) (maior) wodwynde(C15) (maior) anglice wethewynde(C19) erthwynd, wild lili [= 'lilium sylvestre'](C21) (maior) wodebynd(C21) anglice wodbynd or wyld hoppes(C26) (agrestis) wylde lylye(C32)

Volubilis Media
Calystegia sepium (L.) R.Br. Bellbine, 'Larger Bindweed'
anglice honisouke of þe hegge(C12)

Volubilis Minor
Convolvulus arvensis L. Bindweed
Glechoma hederacea L. Ground Ivy
ere terestre, anglice heyhove(B1) anglice heyhove(B2) anglice heihove(B7)

(minor, terrestris) wylde lyly, erþewynde(C1) anglice heyhove(C6) erth-ybe(C29)

Volubilis Terrestris
Convolvulus arvensis L. Bindweed
wilde lylly(C4)

Vulgago
Asarum europaeum L. Asarabacca, Hazelwort
gillofres(C21)
Lit. Alph., p.192; André I, p. 338 & II, p. 277; N221

Vulvago [volvago]
?
cammok [wylde sauge, expuncted](C7) sauag [sic](C21)
Lit. cf. André I, p. 338

Warencia
Rubia tinctorum L. Madder
warence, madir(A3)
warense, madyr(B8)

Warencia Minor
Galium aparine L. Cleavers, Hairif
secundum aliquos clever, heyrev(C7) aftir sum men clyþen(C31)

Xifion
?
voxesfot(B3)
anglice foxesgloven(C12)
Lit. cf. André I, p. 339 (xiphium) & II, p. 278; B III, 97f.

Xilibrium
Mentha longifolia (L.) Huds. Horse-Mint
anglice horsmynt(C36)

Xiridia
Gladiolus L. ssp. Gladden
Iris L. ssp. Iris
anglice gladwyn(C36)

Ygia
Lamium L./*Urtica* L. ssp. Nettleseed
netelseed(C10) nettilsede(C21) netilsede(C24) anglice archangell(C26)

Ypericon
Hypericum perforatum L. St. John's Wort
anglice seint Johanhisgras(A1) Johanhisgras(A3)

Yperisticon
?*Verbena officinalis* L. ? Vervain
anglice verveyne(B1)

Zaracon
Plantago indica L. Fleabane
prosill(C21)
Lit. N222

Zeclea
Polygonum aviculare agg. Knotgrass
centinodi, sparwistung, swynesgresse(C21)
Lit. André I, p. 340 (zeclias)

Zedoarium [zeodoarium, zedeuarium, zedoare, zeduary]
Curcuma zedoaria Rosc. Zedoary, Turmeric
gallice zedeuale(A1) citouales, sedewaal(A3) anglice et gallice zedewale(A6)
setwale(B4)
sedywale(c1) anglice s(t)etewele(C3) seduall(C4) zedewhale(C7) sedewale(C10) setewale(C21)

Zia
?*Senecio vulgaris* L. ? Groundsel
.i. synchon(C39)

Zibber
Rosmarinus officinalis L. Rosemary
rosemari, feld-madre(C21)
Lit. André I, p. 340 (zibbir); B III, 88 (feldmaedere)

Zinziber
Zingiber officinale Rosc. Ginger
gallice gingivre(A1) gyngir(A3) anglice et gallice ginger(A6)

gyngere(B4)
gyngyver(C1) gingivre(C2) anglice gynger(C3)

Zinzilion [zinzeloon]
Potentilla L. ssp. Cinquefoil
anglice five-levedegras(C10) quinfoile(C21)
Lit. Alph., p.198 (zilzizeleon) note 17

Zirungen [zirunge]
?*Borago officinalis* L. ? Borage
borage(C10, C21, C24)

Zobus
?*Orchis* L. ssp. ?'Orchis
ravenleke(C21)

Zubal
Gladiolus, Iris L. ssp. Gladden
gladyn(C21)
Lit. André I, p. 341

Zucarum [zuccara, zuccura]
Saccharum officinarum L. Sugar Cane
gallice sukre(A1) sukre(A3)
sugyr(C1) anglice sugur(C3) sugre(C10) sucre(C21) anglice suger(C37)

Zizannia [zysannia, zizannium]
Agrostemma githago L. Corn Cockle
Lolium temulentum L. Darnel
anglice cokel(A1) cokel(A3) anglice cokil(A6)
cokkel(B3) nel(B4) neele, cokkel(B8)
cokkyl(C1) darnell(C2) anglice popyll(C3) anglice reveray(C5)
popul(C14) cokkel(C35)

INDEX OF VERNACULAR NAMES

INDEX OF BOTANICAL NAMES

Lightning Source UK Ltd.
Milton Keynes UK
UKHW020626210122
397421UK00003BA/136

9 780859 912730